The
Mass Spectrometry
Primer

Michael P. Balogh

Waters
THE SCIENCE OF WHAT'S POSSIBLE.™

Waters Corporation
34 Maple Street
Milford, MA 01757

Library of Congress Control Number: 2009921480

Printed in the USA

February 2009 715001940 VW-FP

List of Figures & Tables

Preface

Information in any form committed to public view must be of high scholarly order. It is also true that once words have been printed the value of the meaning they impart decreases as new understanding takes shape.

This primer covers a wide range of topics related to the most wide spread of modern mass spectrometry practices and answers some frequently asked questions about the use and capabilities of mass spectrometers. Links are also provided to articles for more in-depth reading. The first section examines who uses mass spectrometers, followed by how compounds are ionized in the source to be analyzed by mass spectrometers. A description of the various types of mass spectrometers is followed by a discussion of the important topics of mass accuracy and resolution—or how well we can tell differences between closely related compounds. Chemistry, sample prep, and data handling are considered, as well as the definition of some terms commonly used in the most prevalent forms of MS practice today.

Primers in different forms can be found from a variety of authors and many of them are referenced for further reading in this one. The electronic version of this primer resides on the Waters website displaying a sidebar offering readers the opportunity to comment (see www.waters.com/primers). So what makes this one different is its inherent continually self-validating existence based on its use as it resides on the web. Succeeding versions of both the electronic version and later editions of the printed version will reflect that.

Michael P. Balogh
Principal Scientist, MS Technology Development
Waters Corporation

Table of Contents

Who Uses Mass Spectrometry?

Before considering mass spectrometry (MS), you should consider the type of analyses you perform and the kind of results you expect from them:

- Do you want to analyze large molecules, like proteins and peptides, or acquire small, aqueous-molecule data?
- Do you look for target compounds at a determined level of detail, or do you want to characterize unknown samples?
- Are your current separations robust, or must you develop methods from complex matrixes?
- Do you require unit mass accuracy, (such as 400 MW), or accuracy to 5 ppm, (such as, 400.0125 MW or 2 mDa at mass 400)?
- Must you process hundreds of samples a day? Thousands? Tens of thousands?

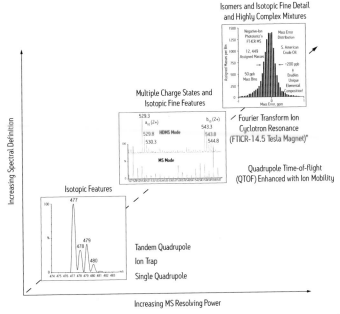

Figure 1: The ability to determine an analyte's character increases with instrument capability.

Purcell, J. M.; Hendrickson, C. L.; Rodgers, R. P.; Marshall, A. G. "Atmospheric Pressure Photoionization Fourier Transform Ion Cyclotron Resonance Mass Spectrometry for Complex Mixture Analysis," Anal. Chem. 2006, 78, 5906-5912.

Researchers and practitioners from various disciplines and sub-disciplines within chemistry, biochemistry, and physics regularly depend on mass spectrometric analysis. Pharmaceutical industry workers involved in drug discovery and development rely on the specificity, dynamic range, and sensitivity of MS to differentiate closely-related metabolites in a complex matrix and, thus, identify and quantify metabolites. Particularly in drug discovery, where compound identification and purity from synthesis and early pharmacokinetics are determined, MS has proved indispensable.

Biochemists expand the use of MS to protein, peptide, and oligonucleotide analysis. Using mass spectrometers, they monitor enzyme reactions, confirm amino acid sequences, and identify large proteins from databases that include samples derived from proteolytic fragments. They also monitor protein folding, carried out by means of hydrogen-deuterium exchange studies, and important protein-ligand complex formation under physiological conditions.

Clinical chemists, too, are adopting MS, replacing the less-certain results of immunoassays for drug testing and neonatal screening. So are food safety and environmental researchers. They and their allied industrial counterparts have turned to MS for some of the same reasons: PAH and PCB analysis, water quality studies, and pesticide residue analysis in foods. Determining oil composition, a complex and costly prospect, fueled the development of some of the earliest mass spectrometers and continues to drive significant advances in the technology.

Today, the MS practitioner can choose among a range of ionization techniques that have become robust and trustworthy on a variety of instruments with demonstrated capabilities.

See MS – The Practical Art, LCGC® (www.chromatographyonline.com)

Profiles in Practice Series: Metabolism ID and Structural Characterization in Drug Discovery, Vol. 23, No. 2, February 2005

Why this is important: Illustrates and contrasts approaches used in metabolite identification practice as described by two leading practitioners.

Profiles in Practice Series: Stewards of Drug Discovery—Developing and Maintaining the Future Drug Candidates, Vol. 23, No. 4, April 2005

Why this is important: Compares developing and handling drug candidate compounds and libraries from the viewpoint of a large pharmaceutical company and a small specialty company.

What are Mass Spectrometers? How Do They Work?

Mass spectrometers can be smaller than a coin, or they can fill very large rooms. Although the various instrument types serve in vastly different applications, they nevertheless share certain operating fundamentals. The unit of measure has become the Dalton (Da) displacing other terms, such as amu. 1 Da = 1/12 of the mass of a single atom of the isotope of carbon 12 (^{12}C).

Once employed strictly as qualitative devices—adjuncts in determining compound identity—mass spectrometers were once considered incapable of rigorous quantitation. But in more recent times, they have proved themselves as both qualitative and quantitative instruments.

A mass spectrometer can measure the mass of a molecule only after it converts the molecule to a gas-phase ion. To do so, it imparts an electrical charge to molecules and converts the resultant flux of electrically-charged ions into a proportional electrical current that a data system then reads. The data system converts the current to digital information, displaying it as a mass spectrum.

Profiles in Practice Series: A Revolution in Clinical Chemistry, Vol. 23, No. 8, August 2005
Why this is important: Health care professionals have recently embraced MS as a means to greatly improving the accuracy, speed, and quality of patient information but it is a work-in-progress.

Profilets in Practice Series: Advances in Science and Geopolitical Issues (Food Safety), Vol.23, No. 10, October 2005
Why this is important: As instruments become more robust and sensitive, MS is changing the ways of regulated testing with far. -reaching global consequences

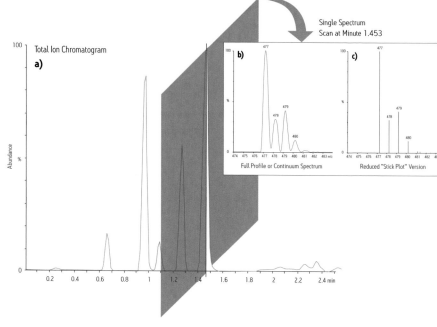

Figure 2: a) Increasing abundance in the total ion current (TIC) is represented as it changes over time in a chromatographic-like trace. b) Each digital slice of a peak represents the ions at that time making up the ion current often referred to as a profile or continuum acquisition. The x or 'time' axis is now the mass-to-charge ratio (m/z) the ability to resolve neighboring ions in the spectrum (such as isotopes) is readily seen. c) A profile spectrum is often reduced to a 'stick plot' represented by centroids dropped from each peak apex reducing the size of the stored file in favor of the increased resolution information.

Ions can be created in a number of ways suited to the target analyte in question:

1) By laser ablation of a compound dissolved in a matrix on a planar surface such as by Matrix-Assisted Laser Desorption Ionization (MALDI).

2) By interaction with an energized particle or electron, such as in electron ionization (EI).

3) A part of the transport process itself as we have come to know electrospray ionization (ESI) where the eluent from a liquid chromatograph receives a high voltage resulting in ions from an aerosol.

The ions are separated, detected, and measured according to their mass-to-charge ratios (m/z). Relative ion current (signal) is plotted versus m/z producing a mass spectrum. Small molecules typically exhibit only a single charge: the m/z is therefore some mass (m) over 1. The '1' being a proton added in the ionization process (represented M+H$^+$ or M-H$^-$ if formed by the loss of a proton) or if the ion is formed by loss of an electron it is represented as the radical cation (M^{+}). The accuracy of a mass spectrometer, or how well it can measure the actual true mass, may vary as will be seen in later sections of this primer.

Larger molecules capture charges in more than one location within their structure. Small peptides typically may have two charges (M+2H$^+$) while very large molecules have numerous sites, allowing simple algorithms to deduce the mass of the ion represented in the spectrum.

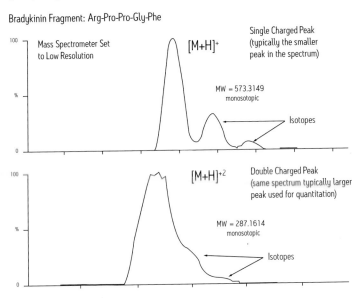

Bradykinin Fragment: Arg-Pro-Pro-Gly-Phe

Mass Spectrometer Set to Low Resolution

[M+H]$^+$

Single Charged Peak (typically the smaller peak in the spectrum)

MW = 573.3149 monosotopic

Isotopes

[M+H]$^{+2}$

Double Charged Peak (same spectrum typically larger peak used for quantitation)

MW = 287.1614 monosotopic

Isotopes

Figure 3: Low resolution instruments can deliver exceptional accurate mass when properly calibrated, but as more data crowds its limited resolution space provides less information about the spectrum. A common metabolic fragment (BK1-5 or Arg-Pro-Pro-Gly-Phe) of Bradykinin, a 9 amino acid peptide, ACE (angiotensin converting enzyme) inhibitor used to dilate blood vessels can carry two charges (single charge or M+H yields monoisotopic value 573.3149 while the doubly charged version or M+2H displays 287.1614). The isotopes are doubly charged as well begin to fill the available resolution space.

How Large a Molecule can I Analyze?

Desorption methods (as described on page 22) have extended the ability to analyze large, nonvolatile, fragile molecules. Routine detection of 40,000 Da within 0.01% accuracy (or within 4 Da) allows the determination of minor changes, such as post-translational modification of proteins. Multiple charging extends the range of the mass spectrometer well beyond its designed upper limit to include masses of 1,000,000 Da or more.

Isotope and Elemental Mass Spectrometry

Natural isotope abundance is well characterized. Though often thought to be stable, it can nevertheless display significant and characteristic variances. Isotope ratio measurements are used in metabolic studies (isotope-enriched elements serve as tracers) and also in climatic studies that measure temperature-dependent oxygen and carbon changes. In practice, complex molecules are reduced to simple molecular components before being measured using high-accuracy capabilities, such as those found on magnetic sector instruments (see the following section).

Elemental analysis is typically performed on inorganic materials—to determine elemental makeup, not structure—in some cases using solid metal samples. Inductively coupled plasma (ICP) sources are common where a discharge (or lower power glow discharge) device ionizes the sample. Detection using dedicated instruments, at the parts-per-trillion level, is not uncommon.

Common Ionization Methods

Electron Ionization (EI)

Many are familiar with electron ionization (EI). (Sometimes the earlier phrase "electron impact" is used—although, technically, it is incorrect.) EI, often performed by exposing a sample to 70 eV electrons, is referred to as a "hard" technique. The energy of the electrons interacting with the molecule of interest is generally much greater than that contained in its bonds, so ionization occurs. The excess energy breaks bonds in a well-characterized way. The result is predictable, identifiable fragments from which we can deduce the molecule's identity. Imparting energy beyond abstraction of only an electron from the outer shell which yields a radical cation in the positive mode (M^{+}) produces a rich spectrum of fragments. Unlike "softer" atmospheric-ionization techniques which produce a spectral response sometimes characteristic of the manufacturer's particular source design, the EI technique is fairly independent of the source design. A spectrum produced by one EI instrument looks much like a spectrum of the same compound from another EI instrument, a fact that lends itself to creating spectral libraries to match unknowns to reference spectra.

Chemical Ionization (CI)

Molecules that fragment excessively call for "soft" techniques. Chemical ionization (CI) produces ions by a gentler proton transfer process that preserves and promotes the appearance of the molecular ion itself. The sample is exposed to an excess of reagent gas such as that which evolves when methane forms the protonated molecular ion (M+H). The reverse process can produce negative ions. Transferring the proton to the gas molecule can, in some cases, produce the negative ion (M-H).

CI is sometimes used for compounds with chemistry similar to those analyzed by EI to enhance the abundance or appearance of the molecular ion in favor of significant fragmentation. Similar to EI, samples must be thermally stable since heating in the source causes vaporization. The ionization mechanism of CI relies on EI for the initial ionization step but within the source is a chemical reagent gas, such as methane, ISO butane, or ammonia, at high pressure. The reagent gas, which is present at a much higher concentration than the analyte (R), is ionized by electron ionization to give primary R^{+t}. reagent ions. The collision of the R^{+}. ions with neutral R molecules lead to the formation of stable secondary ions which are the reactant species which then ionize analyte molecules (A) by ion-molecule reactions.

For example the ion-molecule reaction between a methane ion and a methane molecule gives rise to the fairly stable CH_5^+ species.

$$CH_4^{+\cdot} + CH_4 \quad \blacktriangleright \quad CH_5^+ + CH^3.$$

The reactant ion CH_5^+ can ionize neutral analyte molecules (A) by proton transfer, hydride abstraction, or charge exchange.

$$RH^+ + A \quad \blacktriangleright \quad R + AH^+ \text{ (proton transfer)}$$

$$(R\text{-}H)^+ + A \quad \blacktriangleright \quad R + (A\text{-}H)^+ \text{ (hydride abstraction)}$$

$$R^{+\cdot} + A \quad \blacktriangleright \quad R + A^{+\cdot} \text{ (charge exchange)}$$

The most common ionization reactions are protonation, which is favored for molecules with proton affinities higher than the reagent. Hydride abstraction is common for lower proton affinity molecules and charge exchange occurs with reagents of high ionization energy.

The substance to be analyzed is at a much lower pressure than the reagent gas. If we consider methane as the reagent gas the electron impact causes mainly ionization of the methane. This fragments in part to CH_3^+. These species then undergo ion molecule reactions under the high source pressures employed.

$$CH_4^{+\cdot} + CH_4 \quad \blacktriangleright \quad CH_5^+ + CH^3.$$

$$CH_3^+ + CH_4 \quad \blacktriangleright \quad C_2H_5^+ + H_2$$

CH_5^+ can act as a Bronsted acid and $C_2H_5^+$ as a Lewis acid to produce ions from the analyte.

Careful choice of the CI reagent gas can improve charge transfer to an analyte molecule as the gas phase acidity of the chemical ionization gas influences the efficiency of the charge transfer. In CI the analyte is more likely to result in a molecular ion with the reduced fragmentation conserving the energy normally internalized in EI to break bonds.

Negative Ion Chemical Ionization (NCI)

A variation, negative ion chemical ionization (NCI), can be performed with an analyte that contains electron-capturing moieties (e.g., fluorine atoms or nitrobenzyl groups). Sensitivity can be increased many-fold (reported to be 100 to 1000 times greater in some case) than that of EI. NCI is applicable to a wide variety of small molecules that are (or can be) chemically modified to promote electron capture.

In negative ion there are two primary mechanisms whereby negative ions are produced: electron capture and reactant ion chemical ionization. Under CI conditions electronegative molecules can capture thermal electrons to generate negative ions. True negative ion chemical ionization occurs by reaction of an analyte compound (AH) with negatively charged reactant ions (R^{-} or R^{-}). Several types of ion-molecule reactions can occur, the most common being proton abstraction.

$$AH \; + \; R^{-} \qquad \qquad A^{-} \; + \; RH$$

As the proton affinity (basicity) of the reactant ion increases the more likely proton abstraction is to occur.

Common Separation and Sample Delivery Methods

Gas Chromatography (GC)

Perhaps the first encounter with a mass spectrometer for many is as the detector for a gas chromatograph. The range of GC/MS instrument types has expanded to transcend the limitations of earlier instrument designs or to meet increasingly stringent legislation in applications like environmental analysis, food safety screening, metabolomics, and clinical applications like forensics, toxicology, and drug screening.

In the past, two types of mass spectrometers dominated GC/MS analysis: magnetic sector and the single quadrupole instruments. The former, which offered high resolution and accurate mass analyses, was used in applications that required extreme sensitivity. The latter performed routine analysis of target compounds.

The most challenging GC/MS analyses were reserved for magnetic sector instruments: dioxins in environmental/industrial samples or screening for the illegal use of performance-enhancing drugs in competitive sports. Femtogram detection levels, at high resolution/selectivity, are easily achieved on magnetic sector instruments.

Shortly after their introduction, quadrupole GC/MS systems gained acceptance in target analysis applications. United States Environmental Protection Agency (US EPA) Methods dictated the use of quadrupole GC/MS instruments to analyze samples for numerous environmental contaminants. Because those applications require only picogram-to-nanogram levels of detection, the poorer sensitivity of the quad relative to the sector was not a limitation. On the contrary, the greatly reduced cost, ease-of-use, and portability proved a benefit.

Liquid Chromatography (LC)

The revolutionary technology that gave us analytical access to about 80% of the chemical universe unreachable by GC is also responsible for the phenomenal growth and interest in mass spectrometry in recent decades. A few individuals are singled out (see the section on 'A Brief History' of Mass Spectrometry) for coupling LC with MS. Beginning arguably in the 1970s, LCMS as we know it today reached maturation in the early 1990s. Many of the devices and techniques we use today in practice are drawn directly from that time.

Liquid chromatography was defined in the early 1900's by the work of the Russian botanist, Mikhail S. Tswett. His studies focused on separating leaf pigments extracted from plants using a solvent in a column packed with particles. In its simplest form, liquid chromatography relies on the ability to predict and reproduce, with great precision, competing interactions between analytes in solution (the mobile or condensed phase) being passed over a bed of packed particles (the stationary phase). Development of columns packed with a variety of functional moieties in recent years and the solvent delivery systems able to precisely deliver the mobile phase has enabled LC to become the analytical backbone for many industries. The acronym HPLC was coined by Csaba Horváth in 1970 to indicate that high pressure was used to generate the flow required for liquid chromatography in packed columns. Continued advances in performance since then, including development of smaller particles and greater selectivity, changed the acronym to high performance liquid chromatography.

In 2004, further advances in instrumentation and column technology increased resolution, speed, and sensitivity in liquid chromatography. Columns with smaller particles (1.7 micron) and instrumentation with specialized capabilities designed to deliver mobile phase at 15,000 psi (1,000 bar) came to be known as Ultra Performance Liquid Cheomatography (UPLC® Technology) representing the differentiated term ultra performance liquid chromatography. Much of what is embodied in this current technology was predicted by investigators such as John Knox in the 1970s. Knox predicted the optimum particle diameters would be 1-2 µm and chromatography would be thermally sensitive to frictional heat. Technology capable of developing robust, uniform small particles was necessarily encountered and resolved on the path to developing UPLC Technology for widespread use. A good basic primer on HPLC and UPLC can be seen at www.waters.com/primers.

Atmospheric Ionization Methods

Electrospray Ionization (ESI)

The general term "atmospheric pressure ionization" (API) includes the most notable technique, ESI, which itself provides the basis for various related techniques capable of creating ions at atmospheric pressure rather than in a vacuum (torr). The sample is dissolved in a polar solvent (typically less volatile than that used with GC) and pumped through a stainless steel capillary which carries between 2000 and 4000 V. The liquid aerosolizes as it exits the capillary at atmospheric pressure, the desolvating droplets shedding ions that flow into the mass spectrometer, induced by the combined effects of electrostatic attraction and vacuum.

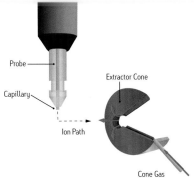

Probe

Capillary

Extractor Cone

Ion Path

Cone Gas

Figure 4: Simplified schematic of an ESI probe positioned in front, and orthogonal to, the MS ion inlet. A cone or counter-current gas is often applied to aid desolvation of liquid droplets as they enter the rarified gas vacuum region of the analyzer.

The mechanism by which potential transfers from the liquid to the analyte, creating ions, remains a topic of controversy. In 1968, Malcolm Dole first proposed the charge-residue mechanism in which he hypothesized that as a droplet evaporates, its charge remains unchanged. The droplet's surface tension, ultimately unable to oppose the repulsive forces from the imposed charge, explodes into many smaller droplets. These Coulombic fissions occur until droplets containing a single analyte ion remain. As the solvent evaporates from the last droplet in the reduction series, a gas-phase ion forms.

In 1976, Iribarne and Thomson proposed a different model, the ion-evaporation mechanism, in which small droplets form by Coulombic fission, similar to the way they form in Dole's model. However, according to ion evaporation theory, the electric field strength at the surface of the droplet is high enough to make leaving the droplet surface and transferring directly into the gas phase energetically favorable for solvated ions.

It is possible that the two mechanisms may actually work in concert: the charge residue mechanism dominant for masses higher than 3000 Da while ion evaporation dominant for-lower masses (see R. Cole, "Some Tenets Pertaining to Electrospray Ionization Mass Spectrometry", Journal of Mass Spectrometry, 35, 763-772 [2000]).

The liquid from the liquid chromatograph enters the ESI probe in a state of charge balance. So when the solvent leaves the ESI probe it carries a net ionic charge. To ensure that ESI is a continuous technique, the solution must be charged by electrochemical reactions whereby electrons transfer to a conductive surface acting as an electrode. Among other effects, this process can lead to pH changes. It is assumed that, in positive mode, positive-charged droplets leave the spray and electrons are accepted by the electrode (oxidation). (The reverse would be true in negative mode.) The surface area of the electro-active electrode, the magnitude of the current, and the nature of the chemical species and their electrode potentials all exert an effect.

Over all, ESI is an efficient process. However, the activation energy and energy difference for the reaction, in total, for individual species varies. The flow rate of the solution and the applied current define limits for each droplet. Competition between molecules occurs, and suppression of analytes of interest is not uncommon.

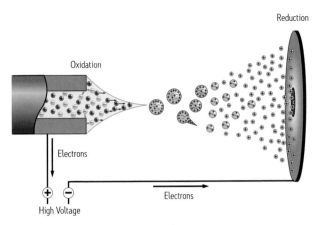

Figure 5: After formation the ions are "dragged" through a potential gradient (an electric field) to the counter plate.*

*Figure courtesy Andreas Dahlin (www.adorgraphics.com)

Extensions of basic ESI theory, such as reducing the liquid to extremely low volumes—for example to 30 nL/min in the case of nanospray—have proved effective, especially in sample-limited studies of proteins and amino acids.

Atmospheric Pressure Chemical Ionization (APCI)

Although work demonstrating Atmospheric Pressure Chemical Ionization (APCI) was published in parallel with that demonstrating ESI, APCI was not widely adopted until ESI was commercialized, which occurred in the wake of Fenn's work in 1985.

Horning first introduced APCI in 1973 to analyze volatile compounds using various introduction techniques, one of which being HPLC. The adjunctive capability of APCI permits analytes that resist conversion to gas-phase ions by ESI, the less polar and more volatile ones introduced into a mass spectrometer from a condensed phase (or liquid) stream. Unlike ESI, APCI transfers neutral analytes into the gas phase by vaporizing the introduced liquid in a heated gas stream. Chemical ionization relies on the transfer of charged species between a reagent ion and a target molecule to produce a target ion that can be mass analyzed. Most commonly, in positive ion mode, an adduct forms between the target molecule and the small H^+ ion, although adducts with salts are common as well. For example, the ammonium adduct can form $(M+NH_4)^+$ when the weak-acid weak-base salt ammonium acetate, a modifier often used in place of the less volatile and highly ionic phosphate buffer, is present in the mobile phase. At higher salt concentrations, competition between the protonated and ammoniated forms can produce a decreased response for both. The maximum number of ions capable of forming by APCI is much greater than in ESI because reagent ions form redundantly. The liquid is pushed through a nonconductive tube, usually of fused-silica glass, around which a nebulizing gas flows. The resultant fine droplets collide with the inner, heated wall of a tube or probe that extends beyond the end of the nonconductive tube, and are thus converted to the gas phase. This ionization type is often performed at much greater linear velocities than flow rates normally associated with electrospray. Contemporary instruments however provide much greater desolvation capacities enhancing performance for all aerosol dependent techniques allowing multi-mode capabilities such as ESCi® (high speed millisecond switching between ESI and APCI in the same ESI source).[1]

The desolvated analyte molecules are then ionized via chemical ionization. The ionizing potential is applied, not through the liquid as in ESI, but at the tip of a needle as a plasma, or corona, through which the droplets pass. In effect, the mobile phase acts as an intermediary transferring the charge to the analyte. Hence the early name given APCI: "solvent-mediated electrospray."

Bio-Molecular Ionization Methods

Ionization techniques have been developed to aid identification of biomolecules rather than aggressively reduce the molecule to components. Two "energy deposition" processes, electron-capture dissociation (ECD)[2] and electron-transfer dissociation (ETD)[3] are commonly recognized in biomolecular analysis and proteomics. Both cleave bonds adjacent to sites of electron capture and, unlike other fragmentation processes, such as collision-induced dissociation (CID), the cleaved bonds are not the most labile within the molecule. The cleavages observed are less dependent on the peptide sequence so cleavages between most amino acids in the peptide backbone tend to be independent of the molecule's size. The dominant fragmentation in ECD and ETD of peptides is the formation of c and z ions. ECD has been demonstrated to be useful for the analysis of labile post-translational modifications, such as phosphorylation and O-glycosylation and for fragmentation analysis of intact proteins.

[1] A Case for Congruent Multiple Ionization Modes in Atmospheric Pressure Ionization Mass Spectrometry, Chapter 5, M.P. Balogh, Journal of Chromatography, Vol. 72, 2007, Advances in LC-MS Instrumentation, ed. Achille Cappiello.
[2] R.A. Zubarev, Electron-capture dissociation tandem mass spectrometry, Curr. Opin. Biotechnol 15 (2004), pp. 12-16
[3] J.J. Coon, J. Shabanowitz, D.F. Hunt and J.E. Syka, Electron transfer dissociation of peptide anions, J. Am. Soc. Mass Spectrom 16 (2005), pp. 880-882

ESI mass spectrometry has been shown to be further aided for elucidating structural details of proteins in solution when coupled with amide hydrogen/deuterium (H/D) exchange analysis. Charge-state distributions and the envelopes of charges ESI forms on proteins can provide information on solution conformations of larger proteins with smaller amounts of sample not easily performed using other techniques, such as near ultraviolet circular dichroism (CD) and tryptophan fluorescence (however it is typically used in conjunction with these techniques and others such as nuclear magnetic resonance). The other techniques measure the average properties of large populations of proteins in solution so an additional advantage seen with MS is its ability to provide structural details on transient or folding intermediates.

Alternative Ionization Means

Pure or neat compounds can be introduced into the ion source having been deposited in the tip of a rod, or solids probe. With heat, the sample sublimates or evaporates into the gas phase. In most cases, ionization follows by the means described here. But in some cases, ionization occurs simultaneous with the sublimation or evaporation.

Atmospheric-Pressure Photo Ionization (APPI)
- Direct or dopant-assisted photon ionization of analytes with ionization potential below 10 eV (the primary photon energy output by a krypton gas lamp). The ionization potential for solvents commonly used in LC are above 10 eV. APPI is one of the primary API alternatives in the lab since it extends the ionization range to more non-polar analytes than either ESI or APCI can ionize.

Matrix-Assisted Laser Desorption (MALDI)
- Soft ionization for intact proteins, peptides, and most other biomolecules (oligonucleotides, carbohydrates, natural products, and lipids) and analysis of heterogeneous samples (analysis of complex biological samples such as proteolytic digests).

- High energy photons interact with a sample embedded in an organic matrix typically with sub-pico mole sensitivity.

- First introduced in 1988 by Tanaka, Karas, and Hillenkamp.

Fast-Atom Bombardment (FAB)
- An early form of soft ionization using a stream of cesium ions to "sputter" ions from a sample dissolved in a glycerol, or similar, matrix.

Desorption

- Plasma Desorption (PD): nuclear fission fragments interact with a solid sample deposited on metal foil.

- Secondary-Ion MS (SIMS): high velocity ions impact a thin film of sample deposited on metal plate or contained in a liquid matrix (liquid SIMS).

- Field Desorption: a high field gradient is imposed on a sample deposited on a support.

- Desorption Electrospray Ionization (DESI): along with closely related techniques like direct analysis real time (DART), atmospheric solids analysis probe (ASAP) and others recently introduced to the market, these tend to create ions secondary to some interaction on a surface. In DESI, an energized liquid stream is aimed at a sample deposited in a flat surface, causing secondary ionization to occur at atmospheric pressures.

See MS – The Practical Art, LCGC (www.chromatographyonline.com)

Incipient Technologies: Desorption and Thermal Desorption Techniques, Vol. 25, No. 10 December 2007
Why this is important: Describes and compares techniques such as DESI, DART, and ASAP with realistic appraisals for their use.

Alternatives in the Face of Chemical Diversity, Vol. 25, No. 4, April 2007
Why this is important: Explores prospects for applying GC and other not-so-typical techniques to contemporary instruments designed for atmospheric work.

Ionization Revisited, Vol. 24, No. 12, December 2006
Why this is important: Provides an overview of the major ionization techniques in use today with references.

Also see:

Balogh, M.P., The Commercialization of LC-MS During 1987-1997: A Review of Ten Successful Years, LC/GC, Vol. 16, No. 2, 135-144, February 1998

Gary J. Van Berkel, Sofie P. Pasilis, and Olga Ovchinnikova, Established and Emerging Atmospheric Pressure Surface Sampling/Ionization Techniques for Mass Spectrometry, Journal of Mass Spectrometry. 2008; 43: 1161-1180, July 2008

A Brief History of Mass Spectrometry

1897 – Modern mass spectrometry (MS) is credited to the cathode-ray-tube experiments of J.J. Thomson of Manchester, England.

1953 – Wolfgang Paul's invention of the quadrupole and quadrupole ion trap earned him the Nobel Prize in physics.

1968 – Malcolm Dole developed contemporary ESI but with little fanfare. Creating an aerosol in a vacuum resulted in a vapor that was considered too difficult to be practical. Liquid can represent a volume increase of 100 to 1000 times its condensed phase (1 mL/min of water at standard conditions would develop 1 L/min of vapor).

1974 – APCI was developed by Horning basedlargely on GC, but APCI was not widely adopted.

1983 – Vestal and Blakely's work with heating a liquid stream became known as thermospray. It became a harbinger of today's commercially applicable instruments.

1984 – Fenn's work with ESI was published, leading to his Nobel Prize-winning work published in 1988.

For more historical detail see **www.masspec.scripps.edu/mshistory**

What Types of Instruments are in Use?

In mass spectrometry, the ability to exercise control over experiments is supremely important. Once an ion is created under carefully-controlled conditions it must be detected as a discrete event with appropriate sensitivity. The minimal vapor load made GC an ideal early choice as a hyphenated technology but for only some 20% of compounds. Today, we most frequently aerosolize LC eluent as the means of introducing analytes for ionizing within a mass spectrometer, a technique that requires a vacuum environment to ensure control.

An important design element of any mass spectrometer is pumping capacity. Vacuum must be well distributed in the rarified atmospheric regions of an instrument, and it must be sufficient enough to offset such design necessities as the size of the ion inlet and the amount of vapor needing removal.

Inlet	Pumping Needed to Maintain Analytical Pressures
Capillary GC (1 µL/min)	~400
Microbore LC (10 µL/min)	~5,000
Conventional column LC (1 mL/min)	~50,000

Table 1: An approximation of pumping capacities needed (L/sec) to remove resulting vapor and maintain typical analytical pressures of 3×10^{-6} torr (4×10^{-6} mbar) to detect ions as discrete events relative to the inlet used.

LC/MS pumping requirements depend on the interface used. Ultimately this was one of the reasons spurring development of the API source where the vapor is removed before entry to the MS.

The Analyzer: The Heart of a Mass Spectrometer

The analyzer is an instrument's means of separating or differentiating introduced ions. Both positive and negative ions, as well as uncharged, neutral species, form in the ion source. However, only one polarity is recorded at a given moment. Modern instruments can switch polarities in milliseconds, yielding high fidelity records even of fast, transient events like those typical of UPLC or GC separations in which peaks are only about one second wide.

Quadrupoles and Magnetic Sectors

In 1953, the West German physicists Wolfgang Paul and Helmut Steinwedel described the development of a quadrupole mass spectrometer. Superimposed radio frequency (RF) and constant direct current (DC) potentials between four parallel rods were shown to act as a mass separator, or filter, where only ions within a particular mass range, exhibiting oscillations of constant amplitude, could collect at the analyzer.

Manufacturers of today's instruments target them for specific applications. Single quadrupole mass spectrometers require a clean matrix to avoid the interference of unwanted ions, and they exhibit very good sensitivity.

Triple quadrupoles, or tandem (see Quadrupoles), mass spectrometers (MS/MS) add to a single quadrupole instrument an additional quadrupole, to a single quadrupole instrument which can act in various ways. One way is simply to separate and detect the ions of interest in a complex mixture by the ions' unique mass-to-charge (m/z) ratio. Another way that an additional quadrupole proves useful is when used in conjunction with controlled fragmentation experiments. Such experiments involve colliding ions of interest with another molecule (typically a gas like argon). In such an application, a precursor ion fragments into product ions, and the MS/MS instrument identifies the compound of interest by its unique constituent parts.

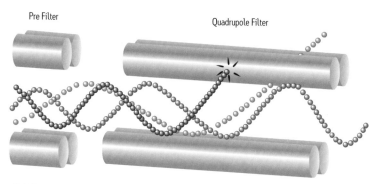

Pre Filter

Quadrupole Filter

Figure 6: When set to filter a specific ion other masses will be lost through a variety of means such as impacting the quadrupole rods themselves or simply leaving the path to the detector.

The quadrupole analyzer consists of four rods, which are usually arranged in parallel and made of metal, such as molybdenum alloys. A tremendous amount of art and science has been invested in developing quadrupole design. Masses are sorted by the motion of their ions, which DC and RF fields induce into an instrument's analyzer. Systematically changing the field strength via the operating software in effect alters which m/z value is filtered or transmitted through to the detector at any given time. Quadrupoles yield a lower resolution than some mass spectrometer designs, such as time-of-flight (TOF) instruments. Yet quadrupoles are relatively simple, easy-to-use, and highly utilitarian instruments that offer a variety of interfaces at a relatively low cost.

Some terminology, more fully defined later in this primer, is necessary for comparing and describing MS capabilities:

Resolving power (often abbreviated as "res" - the ability of a mass spectrometer to separate two masses:

- Low res = unit mass = 1000

- Higher or moderate res = 1000 to 10,000

- High res =10,000+

- Very high res = as much as 3 - 5 million

A more detailed examination of resolution and how we measure it appears in the section "Mass Accuracy and Resolution." Exact mass is the theoretical exact value for the mass of a compound whereas accurate mass is the measured mass value for a compound with an associated error bar like 5ppm. Accurate mass is also commonly used to refer to the technique rather than the measured mass.

MS/MS – Describes a variety of experiments (multiple reaction monitoring [MRM] and single-reaction monitoring [SRM]) that monitor the transition of precursor ions, or fragmentations, to product ion(s), which in general tend to improve the selectivity, specificity, and/or sensitivity of detection over a single-stage-instrument experiment. Two mass analyzers in series, or two stages of mass analysis, in a single instrument are used.

In a triple quadrupole mass spectrometer, there are three sets of quadrupole filters, although only the first and third function as mass analyzers. More recent designs have sufficiently differentiated the middle device (replacing the quadrupole of earlier designs) adding increased function so the term or tandem quadrupole is often used instead. The first quadrupole (Q1), acting as a mass filter, transmits and accelerates a selected ion towards Q2, which is called a collision cell. Although in some designs Q2 is similar to the other two quadrupoles, RF is imposed on it only for transmission, not mass selection. The pressure in Q2 is higher, and the ions collide with neutral gas in the collision cell. The result is fragmentation by CID. The fragments are then accelerated into Q3, another scanning mass filter, which sorts them before they enter a detector.

Fragmentation

CID, also referred to as collisionally-activated dissociation (CAD), is a mechanism by which molecular ions tare fragmented in the gas phase usually by acceleration by electrical potential to a high kinetic energy in the vacuum region followed by collision with neutral gas molecules, such as helium, nitrogen, or argon. A portion of the kinetic energy is converted (or internalized) by the collision which results in chemical bonds breaking and the molecular ion being reduced to smaller fragments. Some similar 'special purpose' fragmentation methods include electron transfer dissociation (ETD), and electron-capture dissociation (ECD). See the section on "Bio-Molecular Ionization Methods".

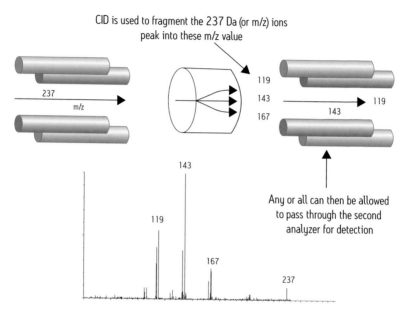

CID is used to fragment the 237 Da (or m/z) ions peak into these m/z value

237

m/z

119
143
167

119

143

Any or all can then be allowed to pass through the second analyzer for detection

143

119

167

237

Figure 7: Endosulfan-ß Product Ion Spectrum. The 237-Da precursor ion entering on the left was fragmented in the MS/MS collision cell. The data system can display only fragments of interest (not all fragments produced) yielding a relatively simple spectrum with respect to the full scan MS spectrum. You can control the extent of fragmentation as you can the choice of precursor ion.

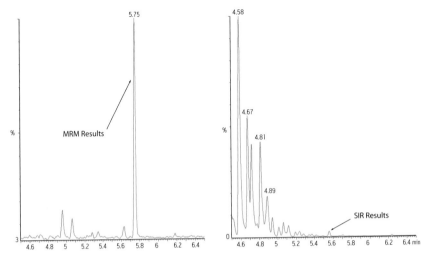

Figure 8: The figure comparing MRM response (left) with SIR response (right) demonstrates how the analyte peak, even when present in solution, may not be determined from SIR data due to chemical background from the matrix. The same GC/MS/MS instrument was used to filter the 146 m/z butylate ion as a precursor, fragment it to product ions (57 m/z shown) to positively, quantifiably identify its presence.

In some regulated industries, to meet the specification for positive compound identification, MRM transitions count for 1.5 "identification points," whereas SIR traces count for 1.0. So, assuming sufficient selectivity, to achieve 3 "IPS," you need 2 MRM transitions but three SIR traces.

Magnetic sector, or a sector field mass analyzer, is an early instrument design that persists today, albeit minimally (having been displaced by modern ESI instruments that can operate in the ESI ion-ization mode). The Waters AutoSpec™, for instance, is used universally for extremely high sensitivity dioxin analysis.

Sectors bend the arc-shaped ion trajectories. The ions' "momentum-to-charge" ratios determine the radius of the trajectories, which themselves are determined by an electric and/or magnetic field. Ions with larger m/z ratios proceed through longer paths than those with smaller ones. The paths are controlled by varying the strength of the magnetic field. Double-focusing mass spectrometers combine magnetic and electric fields in various combinations, although the electric sector followed by the magnetic is more common. This earliest of hybridizations uses the electric sector to focus ions by their kinetic energy as they exit the source. Angular focusing preceded by energetic focusing yields separations of ions with the same nominal mass but different chemical formulas.

Ion Traps and Other Non-Scanning Instruments

An ion trap instrument operates on principles similar to those of a quadrupole instrument. Unlike the quadrupole instrument, however, which filters streaming ions, both the ion trap and more capable ion cyclotron (ICR) instrument store ions in a three-dimensional space. Before saturation occurs, the trap or cyclotron allows selected ions to be ejected, according to their masses, for detection. A series of experiments can be performed within the confines of the trap, fragmenting an ion of interest to better define the precursor by its fragments. Fields generated by RF voltages applied to a stacked or "sandwich" geometry (end-cap electrodes at opposing ends) trap ions in space between the two electrodes. Ramping or scanning the RF voltage ejects ions from their secular frequency, or trapped condition. Dynamic range is sometimes limited. The finite volume and capacity for ions limits the instrument's range, especially for samples in complex matrices.

Ion trap instruments were introduced in the 1980s, but limitations imposed by the internal ionization scheme used in those early instruments prevented their use for many applications. Only with the advent of external ionization did the instruments become more universally practical. The ability to perform sequential fragmentation and, thus, derive more structural information from a single analyte (i.e., fragmenting an ion, selecting a particular fragment, and repeating the process) is called MSn. GC chromatographic peaks are not wide enough to allow more than a single fragmentation (MS/MS or MS^2). Ion trap instruments perform MS/MS or fragmentation experiments in time rather than in space, like quadrupole and sector instruments, so they cannot be used in certain MS/MS experiments like neutral loss and precursor ion comparisons. Also, in MS/MS operation with an ion trap instrument, the bottom third of the MS/MS spectrum is lost, a consequence of trap design. To counter the loss, some manufacturers make available, via their software, wider scan requirements that necessitate the switching of operating parameters during data acquisition.

The trap design places an upper limit on the ratio between a precursor's mass-to-charge ratio (m/z) and the lowest trapped fragment ion, commonly known as the "one-third rule". For example, fragment ions from an ion at m/z 1500 will not be detected below m/z 500—a significant limitation for the de novo sequencing of peptides. The ion trap has limited dynamic range, the result of space-charge effects when too many ions enter the trapping space. Manufacturers have developed automated scanning, which counts ions before they enter the trap, limiting, or gating, the number allowed in. Difficulty can still be encountered when a relatively small amount of an ion of interest is present in a large population of background ions.

Because of similarities in functional design, quadrupole instruments are hybridized to incorporate the advantages of streaming quadrupole and ion trapping behavior to improve sensitivity and allow on-the-fly experiments not possible with either alone. Such instruments are sometimes called linear traps (or Q-traps). The increased volume of a linear trap instrument (over a three-dimensional ion trap) improves dynamic range.

Ion trap instruments do not scan like quadrupole instrument, so using the single ion monitoring (SIM), or single ion recording (SIR), technique does not improve sensitivity on ion traps as it does on quadrupole and s ector instruments.

Fast-fourier transform ion cyclotrons (FTICR) represent the extreme capability of measuring mass with the ability to resolve closely-related masses. Although impractical for most applications, a 14.5-tesla magnet can achieve a resolution of more than 3.5 million and, thus, display the difference between molecular entities whose masses vary by less than the mass of a single electron.

Cyclotron instruments trap ions electrostatically in a cell using a constant magnetic field. Pulses of RF voltage create orbital ionic motion, and the orbiting ions generate a small signal at the detection plates of the cell (the ion's orbital frequency). The frequency is inversely related to the ions' m/z, and the signal intensity is proportional to the number of ions of the same m/z in the cell. At very low cell pressures, a cyclotron instrument can maintain an ion's orbit can for extended periods providing very high resolution measurements.

Sustained off-resonance, irradiation (SORI), is a CID technique used in Fourier-transform ion cyclotron resonance mass spectrometry. The ions are accelerated in cyclotron motion where increasing pressure results in collisions that produce fragments. After the fragmentation, the pressure is reduced and the high vacuum restored to analyze the fragment ions.

TOF instruments, although developed many years ago, have become the basis for much modern work because of their fast, precise electronics and modern ionization techniques, like ESI. A TOF instrument provides accurate mass measurement to within a few parts-per-million (ppm) of a molecule's true mass. A temporally dispersive mass analyzer, the TOF instrument is used in a linear fashion or, aided by electrostatic grids and lenses, as a reflectron. When operated as a reflectron, resolution is increased without dramatically losing sensitivity or needing to increase the size of the flight (or drift) tube.

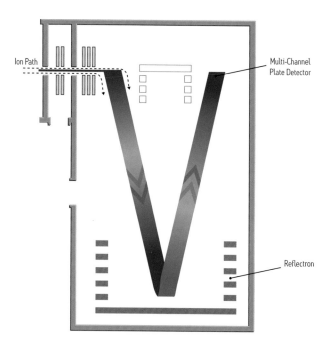

Figure 9: Ions are accelerated by a high-voltage pulse into a drift or flight tube. Lighter ions arrive at the multi-channel plate (MCP or detector) sooner than heavy ones.

TOF analyses involve accelerating a group of ions, in a brief burst, to a detector. The ions exit the source, each having received from a "pusher" electrode an identical electrical charge, or potential. Each ion's potential accelerates, or fires, it into a very low pressure tube. Because all similarly-charged ions share the same kinetic energy (kinetic energy = $mv2$ where m is the ion mass and v the velocity), those with lower masses evidence greater velocity and a lesser interval before striking the detector. Since mass, charge, and kinetic energy determine the arrival time of an ion at the detector, the ion's velocity can be represented as $v = d/t = (2KE/m)^{1/2}$. The ions travel a given distance (d), in time (t), where t depends on the mass-to-charge ratio (m/z). Since all masses are measured for each "push," the TOF instrument can achieve a very high sensitivity relative to scanning instruments.

Today, quadrupole MS systems scan routinely at 10,000 Da (or amu) per second. So a comprehensive scan, even one of short duration (an LC or GC peak of 1 second, for instance) would, nevertheless, capture each ion 10 times, or more, in each second. The TOF instrument's detector registers ions bombarding the plate within nanoseconds of each other. Such resolution offers the added capabilities of a wide dynamic range and greater sensitivity when compared directly to a scanning instrument, such as a quadrupole. Yet the quadrupole instrument is, generally speaking, more sensitive when detecting target analytes in complex mixtures and is, therefore, typically a better quantitation tool. Some instruments, like ion traps, offer a combination of these capabilities, but until the advent of hybrid instruments, no single one could deliver high-order performance in all aspects.

Early MALDI-TOF designs accelerated the ions out of the ionization source immediately. Their resolution was relatively poor and their accuracy limited. Delayed extraction (DE), developed for MALDI-TOF instruments, "cools" and focuses the ions for approximately 150 nanoseconds after they form. Then it accelerates the ions into the flight tube. The cooled ions have a lower kinetic-energy distribution than uncooled ones, and they ultimately reduce the temporal spread of the ions as they enter the TOF analyzer, resulting in increased resolution and accuracy. DE is significantly less advantageous with macromolecules (for instance, proteins >30,000 Da).

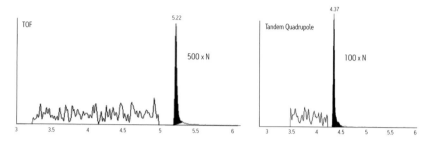

Figure 10: A many fold sensitivity advantage of a TOF over a tandem quadrupole when operating in scan mode can sometimes be seen given the proper conditions (lack of matrix interference for instance) since the TOF does not 'scan' sacrificing duty cycle.

Hybrids

The term "hybrid" applies to various mass spectrometer designs that are composites of existing technologies, such as double-focusing, magnetic sectors and, more recently, ion traps that "front" cyclotrons. One of the most interesting designs, the quadrupole time-of-flight (QTOF) mass spectrometer, couples a TOF instrument with a quadrupole instrument. This pairing results in the best combination of several performance characteristics: accurate mass measurement, the ability to carry out fragmentation experiments, and high quality quantitation.

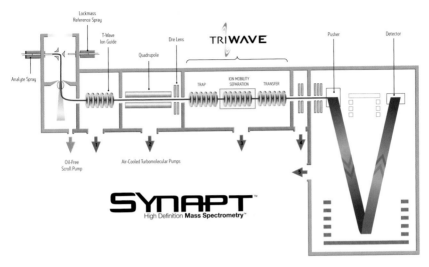

Figure 11: SYNAPT High Definition Mass Spectrometer with TriWave.

Further evolution has produced the coupling of ion mobility measurements and separations with tandem mass spectrometry. Ion mobility mass spectrometry ('IMMS' is used as an acronym here since imaging mass spectrometry is often abbreviated 'IMS') is a technique that differentiates ions based on a combination of factors: their size, shape and charge, and their mass. IMMS devices are commonly used in airports and hand-held field units for rapidly (20 sec) detecting small molecules whose mobility is known: for example certain narcotics and explosives. When adapted to the higher-order instruments, IMMS provides an orthogonal dimension of separation (to both LC and MS) and some unique, enabling capabilities including these:

- Separation of isomers, isobars, and conformers (from proteins to small molecules) and determination of their average rotational collision cross section

- Enhanced separation of complex mixtures (by MS or LC/MS) leading to increased peak capacity and sample clean-up (physical separation of ions, especially chemical noise, and ions that interfere with analytes of interest)

- Performance of CID/IMMS, IMMS/CID, or CID/IMMS/CID and enhancement of the amount of meaningful information that can be gained from fragmentation experiments in structural elucidation studies

In all three analytical scenarios, the combination of high-efficiency ion mobility and tandem mass spectrometry can help overcome analytical challenges that could not be addressed by other analytical means, including conventional mass spectrometry or liquid chromatography instrumentation.

The review article by H.H. Hill Jr., et al., cited at the end of this section, compares and contrasts various types of ion mobility (mass spectrometers available as of the article's 2007 publication) and describes the advantages of applying them to a wide range of analytes. It targets four methods of ion mobility separation currently used with mass spectrometry:

- Drift-time ion mobility spectrometry (DTIMS)
- Aspiration ion mobility spectrometry (AIMS)
- Differential-mobility spectrometry (DMS), also called field-asymmetric waveform ion mobility spectrometry (FAIMS)
- Traveling-wave ion mobility spectrometry (TWIMS)

According to the authors, "DTIMS provides the highest IMS resolving power and it is the only (IMMS) method that can directly measure collision cross-sections. AIMS is a low resolution mobility separation method, but it can monitor ions continuously. DMS and FAIMS offer continuous-ion monitoring capability as well as orthogonal ion mobility separations in which high-separation selectivity can be achieved. TWIMS is a novel (IMMS) method whose resolving power is relatively low. Nevertheless, it demonstrates good sensitivity and is well integrated into operation of a commercial mass spectrometer."

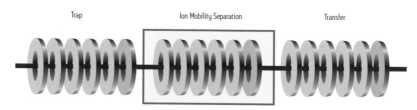

Figure 12: Undifferentiated ions of differing mobility, represented by colored balls, are being trapped, accumulated and released into the T-wave ion mobility separation (IMS) device.

Figure 13: Once released into the T-wave region, a traveling waveform drives the ions through a neutral buffer gas (typically Nitrogen at 0.5 mbar) separating them by their mobility.

Also see:

Special Feature Perspective: Ion Mobility Mass Spectrometry, A. B. Kanu, P. Dwivedi, M. Tam, L. Matz and H. H. Hill Jr., J. Mass Spectrom. 2008; 43: 1-22 Published online in Wiley InterScience, (www.interscience.wiley.com) DOI: 10.1002/jms.1383

Why this is important: A concise overview of ion mobility coupled with MS. One hundred and sixty references on ion mobility mass spectrometry (IMMS) are provided.

An investigation of the mobility separation of some peptide and protein ions using a new hybrid quadrupole/traveling wave IMS/oa-ToF instrument, S. D. Pringle, K. Giles, J. L. Wildgoose, J. P. Williams, S. E. Slade, K. Thalassinos, R. H. Bateman, M. T. Bowers, J. H. Scrivens, Published online (www.sciencedirect.com), International Journal of Mass Spectrometry (2006), doi:10.1016/j.ijms.2006.07.021

Why this is important: Describes how IMMS works with biomolecules.

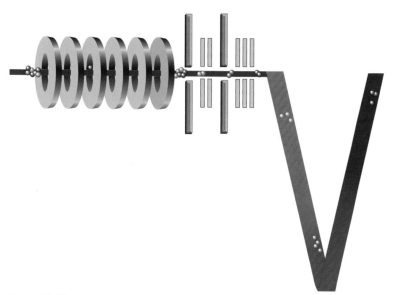

Figure 14: The separated 'packets' of ions with the same mobility characteristics are then passed to the TOF drift tube where their m/z values are measured. The system, therefore, has the potential to separate isobaric ions (ions of identical m/z) or those of very similar m/z prior to mass analysis increasing the overall peak capacity of the MS or LC/MS system.

See MS – The Practical Art, LCGC (www.chromatographyonline.com)

Profiles in Practice Series: The High Speed State of Information and Data Management, Vol. 23 No. 6, June 2005

Why this is important: As data output becomes more complex and voluminous, archiving and retrieving and structured storage emerge as critical issues.

Hardware and software challenges for the near future: Structure elucidation concepts via hyphenated chromatographic techniques, Vol. 26, No. 2, February 2008

Why this is important: The amount of data being developed by modern experiments, which often include MS and orthogonal or hyphenated analytical systems, is discussed.

Coupled with MS, ion mobility is also being applied to investigate the gas-phase structures of biomolecules. Pringle et al. (cited here) examine the mobility separation of some peptide and protein ions using a hybrid quadrupole/traveling wave ion mobility separator/orthogonal acceleration TOF instrument. Comparing mobility data obtained from the traveling wave (TWIMS) separation device with that obtained using various other mobility separators indicates that "while the mobility characteristics are similar, the new hybrid instrument geometry provides mobility separation without compromising the base sensitivity of the mass spectrometer. This capability facilitates mobility studies of samples at analytically significant levels."

Data Handling

You can sum the intensities of ions, and plot them as a function of time (chromatographic retention time) for a total ion chromatogram (TIC) which looks much like the output of a spectrophotometer such as a UV detector. In the case of MS, one axis represents ion intensity; the other can be time or the digital sample taken at a particular time (i.e., a spectrum). You can display each of the spectra can separately, much like a series of images acquired by modern digital video cameras that are, in essence, a series of high speed still photos.

Simple but very useful techniques are possible, for example, reducing the array of data in a selected ion chromatogram or applying digital filters to reduce noise, as you could by displaying only the most intense peak of each digital sample (a base peak ion chromatogram, or BPI).

Data Output, Storage, and Retrieval

Software design has become a separate specialty over the years, not simply a means to set acquisition parameters. Today, operating and data systems permit intricate control of an instrument by its operator.

Significantly, these specialty softtware packages have evolved:

- Workflow controls such as open access (OA), also called "walk-up systems,"—a fully trained operator can make complete LC or GC/MS methods available to a large number of non-specialist users giving them access to advanced technology without the requirement for extensive training. A non-specialist may only need to make occasional use of an instrument for determining a compound's identity or purity. The system allows them access without first becoming proficient operators themselves.

- Data reduction applications — These packages, for instance, may help identify metabolites or develop biomarkers in complex mixtures from the thousands of unique chemical entities. The applications are often augmented by "expert" systems such as principal component analysis software (PCA), which examines trends not otherwise visible in the extensive output.

The demands of data management are fast outstripping the ability to meet them. High resolution, mass-accurate data can generate a prodigious 1 GB/h. Such enormous quantities of data are generated not only by life science investigators but, increasingly, by those working in industries that depend on high volume processes like characterizing the presence of metabolites and their biotransformations. After 180 days of operation, five mass spectrometers, each producing 24 GB of data per day, will present you with the need to store, retrieve, sort, and otherwise make sense of 21.6 terabytes (TB).

The first question in any data scenario must address is how will the data that is collected by used? Unlike e-mail, which imparts its message and thereafter serves little purpose, online data increases in value over time as biological, pharmaceutical, and physicochemical measurements continue to amass within a data file. But this increase in value comes with the cost of ensuring the data's accessibility. In view of the increasing size of data files, and the length of time over which they must be accessed, a solution might include some form of hierarchical storage management. Thus, some smaller percentage of the data would be immediately accessible, or "active," while the remainder, in successive stages, is in-process or earmarked for long-term archiving.

Mass Accuracy and Resolution

Increased, measured mass accuracy and resolution are now dominant tools for structural characterization in various applications beyond early drug discovery. With their broad reach of specificity and utility, QTOF instruments are replacing other LCMS technologies.

Although higher order instruments exist, a QTOF instrument's high mass accuracy falls within a few parts per million of the true, calculated, monoisotopic value, and its high resolution— as much as 10 times higher than a quadrupole instrument's—permits us to determine empirical formulas according to mass defect (where the critical mass value of hydrogen and other atoms present serve as a differentiator). Speciation analysis, discerning the difference between an aldehyde and a sulphide, for example, becomes possible with an increase in mass accuracy above the quadrupole limits to 30 ppm, where the two masses differ by 0.035 Da.

Differentiation between the metabolic processes involving methylation is more demanding, however. Adding CH_2 produces an increase over the precursor (response for the drug alone) in the measured mass of +14.0157 Da, as compared with a two-stage biotransformation involving hydroxylation (addition of oxygen) followed by oxidation at a double bond (loss of H_2), which produces an increase of +13.9792 Da. Yet both measurements, when limited by nominal resolution (a typical quadrupole response) will look like +14 Da.

High Mass Accuracy and Low Resolution

Low resolution quadrupole instruments perform well for extremely high mass-accuracy measurements, like those used for analyzing proteins. The masses of proteins are generally defined as "average" values when the isotope peaks are not resolved relative to each other. Average mass is the weighted mean of all the isotopic species in a molecule. The instrumental resolution normally employed on quadrupole instruments broadens the resolved response for a 10 kDa protein by a factor of x1.27. That factor increases significantly as the mass increases (for example, to x2.65 at 100 kDa). However, by reducing the peak width to m/z 0.25 (increasing resolution to 4000 resolution) rather than limiting the instrument resolution to 1000 using the typical peak width (m/z 0.6) improves the situation dramatically.

In practice, ESI-MS analyses of large molecules produce multiply-charged ions. Hence the widths need to be divided by the number of charges on an ion to give the width on the mass-to-charge ratio scale.

For example, a 20 kDa protein with 10 or 20 charges on it will produce isotope envelopes that are 0.9 or 0.45 m/z units wide at m/z ~2000 or ~1000, respectively.

When these ions are observed on an instrument set for a significantly lower resolution than that required to resolve the isotopes (less than 10,000 resolution), a single peak is produced for each charge state. The overall width is determined by combining the instrument peak width with the theoretical width of the isotopic envelope divided by the number of charges on the ion. The instrumental peak width would be determined on the first isotope peak of a low-molecular-weight compound at the same m/z value as the multiply charged protein peak.

How Much Accuracy Do We Need, or Can Realistically Achieve, and What are the Compromises?

Consider the requirements for unambiguous characterization from the Journal of the American Society for Mass Spectrometry author's guidelines (March 2004). For C, H, O, N compositions (C_{0-100}, H_{3-74}, O_{0-4} and N_{0-4}) a nominal mass-to-charge response at 118 needs only an error not exceeding 34 ppm to be unambiguous, where a m/z response at 750 requires precision better than 0.018 ppm to eliminate "all extraneous possibilities."

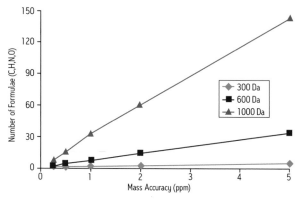

Figure 15: The effect of increasing mass accuracy for unambiguous identification of compounds. *

(Quenzer, T.L., Robinson, J.M., Bolanios, B., Milgram, E. and Greig, M.J., Automated accurate mass analysis using FTICR mass spectrometry, Proceedings of the 50th Annual Conference on Mass Spectrometry and Allied Topics, Orlando, FL, 2002).

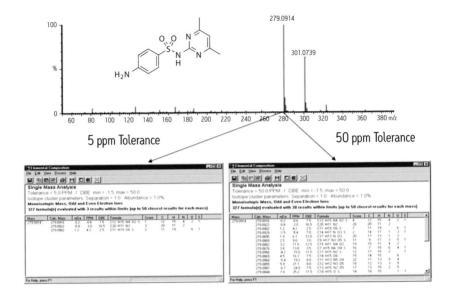

Figure 16: Increased filtering or restriction of error in the measurement reduces the possible candidates for a given result.

Comparing Precision from Instrument to Instrument: Millimass Units (mmu), Measurement Error (ppm), and Resolution

According to the Accurate Mass Best Practice Guide of the VIMMS Program (an initiative that forms part of the UK National Measurement System) most instruments used for accurate mass measurements are capable of achieving precision of 10 ppm or better.

A calculated mass of 118 Da measured by a modern mass spectrometer to within 2 mmu accuracy would display 17 ppm error, sufficient by today's standards for unambiguous determination of a chemical formula of that mass:

Monoisotopic calculated exact mass	= 118 Da
Measured accurate mass	= 118.002 Da
Difference	= 0.002 mmu
Error [Difference/exact mass x 10⁶]	= 17 ppm

An instrument capable of a response at 750 m/z, also deficient by 2 mmu, would be accurate to 2.7 ppm. In the first case, the measurement is more than sufficient for unambiguous identification of a chemical formula, according to the published standards of the Journal of The American Society for Mass Spectrometry. But in the latter case, the measurement is insufficiently precise. Only the highest order Fourier transform ion cyclotron resonance mass spectrometry (FTICR) can achieve such precision at higher masses.

A comprehensive method of evaluating instrument mass accuracy measurement capability which resembles intended use is by calculating the root mean square or Root-Mean Square Error Measurement (RMS) error. To illustrate its use, the following is adapted from the mass measurement accuracy specification of a commercial TOF mass spectrometer.

"The mass measurement accuracy of the instrument, under normal operating conditions, will be better than a given ppm RMS over the given m/z range, based on a number of consecutive repeat measurements of an analyte peak (of given m/z), using a suitable reference peak (of given m/z). Analyte and reference peaks must have sufficient intensity and be free of interference from other masses."

There are some important points and assumptions need to be considered:

1. An **instrument** calibration has already been performed with peaks of known mass using a calibration standard. The reference peak is used to account for any variation in the instrument calibration over time and mass measurement accuracy is determined using the analyte peak.

2. **Normal operating conditions** can also include details of chromatographic conditions (for LC/MS performance specifications) and any related MS operating conditions (e.g. mass resolution, m/z of interest, or spectral acquisition rate).

3. **Sufficient intensity** assumes that the ion count is not detrimental to characterization of the (mass measurement) accuracy and precision of the instrument in question. Too few ions leads to poor ion statistics and too many ions can lead to detector saturation, both of which result in a greater variation in the standard deviation of repeat measurements and will adversely influence calculation of the RMS error (also relevant to instrument calibration).

See MS – The Practical Art, LCGC (www.chromatographyonline.com)

Debating Resolution and Mass Accuracy, Vol. 22 No. 2, 118-130, February 2004

Why this is important: Deals with many of the issues at the heart of practical aspects of use as well as comparisons from industry sources.

Petroleomics: MS from the ocean floor, Vol. 26, No. 3, March 2008

Why this is important: Discusses the utility of the highest resolution instruments and compares them to realistic expectations.

Also see

Methodology for Accurate Mass Measurement of Small Molecules, K. Webb, T. Bristow, M. Sargent, B. Stein, Department of Trade and Industry's VIMMS Program within the UK National Measurement System, (LGC Limited, Teddington, UK 2004). VIMMS 2004 guidance document

Why this is important: A brief and concise overview of issues critical to success with measured, accurate-mass work.

Dealing with the Masses: A Tutorial on Accurate Masses, Mass Uncertainties, and Mass Defects, A. D. Leslie and D. A Volmer, Spectroscopy, Vol. 22, No. 6, June 2007

Why this is important: Expands on the basic topic to include Kendrick mass defect and labeling for peptides and proteins.

4. **Free from interferences** - assumes that the mass measurement of the peak of known mass is free from interference by ions of the same or similar mass. Overlapping peaks lead to poor mass measurement accuracy which is also detrimental to properly characterizing the accuracy or precision of the instrument (also relevant to instrument calibration).

5. **The reference is a good representation** of the m/z range which is relevant to the analysis of a particular sample type.

The RMS error is calculated using the following relation, where E_{ppm} is the ppm error, and n is the number of masses considered:

$$RMS = \sqrt{\frac{\sum (E_{ppm})^2}{n}}$$

It is worth noting the RMS error allows some measurements to fall outside the ppm error "window of interest" (e.g., 5 ppm RMS). To ensure quality measurements, the conditions described above need to be satisfied (particularly regarding intensity and influence of interferences—balanced ion statistics with clear peak definition in the spectra) over a number of repeat injections. Many reported resolution and mass accuracy numbers that you see are not RMS error numbers but instead originate from a single selected (favorable) ion.

It is important to remember in all applications that a weak signal can yield poor ion statistics and can, therefore, be unusable. Too strong a signal can be equally useless, causing detector saturation. Ideally balanced ion statistics with definition in the spectra is the goal.

Some comparisons

With respect to Figure 17:

- Quadrupole resolution is not sufficient to differentiate the two compounds.

- With a resolving power of ~5000, TOF data clearly shows two distinct peaks, which can be accurately mass measured to <5 ppm.

It is important to appreciate the various inter-related roles played in accurate-mass precision by the shift between definitions of mass and increasing resolution and factors such as peak shape and the need for calibration. If these are not clearly understood and taken into consideration mass mis-assignments and other undesirable results may occur.

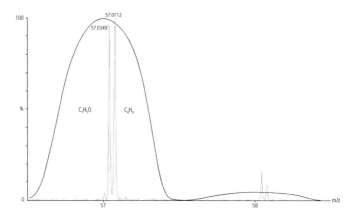

Figure 17: The overlay figure shows quadrupole and TOF response where both mass values on the TOF data are within 1mDa of the exact mass.

The two fragments of different compositions in the figure are from the same analyte and, therefore, in the source at the same time. Even the best chromatography won't help in this case so this emphasizes one of the reasons higher resolution is useful especially in the analysis of unknowns. This applies equally well to QTOF product ion data versus product ion data from a triple quad. As an added advantage with this higher degree of resolution the extracted ion current (XIC) plot of each allows differentiation of the oxygen containing and alkyl containing analytes selectively from the chromatograms where the quad data would lack this capability.

Terminology

Nominal – Unit mass

Sulphamethazine $[C_{12}H_{14}N_4O_2S]$	Nominal = 278

Average mass – Calculated using all isotopes of each element and their natural abundance

Sulphamethazine $[C_{12}H_{14}N_4O_2S]$	Average mass = 278.3313

Calculated exact mass – (Monoisotopic). Determined by summing the masses of the individual isotopes for a given ion

Sulphamethazine $[C_{12}H_{14}N_4O_2S]$	Exact mass = 278.0837

Accurate mass – (Actually "measured exact mass") It is the measure of an m/z reported to (typically) three or four decimal places

As mass increases, differences between the definitions increase, and peak shape plays a bigger role:

Ubiquitin	Nominal	Exact	Average
$[C_{378}H_{630}N_{105}O_{118}S]$	8556	8560.6254	8565.8730

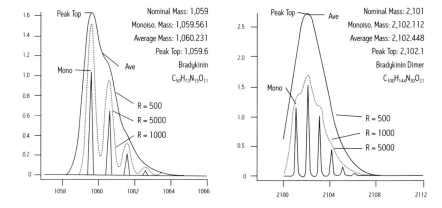

Figure 18: Resolution becomes increasingly important to properly determine the monoisotopic and average mass relative to the peak top as mass increases.[*]

[*] *Unit 16.1, An Overview of Peptide and Protein Analysis by Mass Spectrometry, S. Carr and R. Annan, in Current Protocols in Protein Science, J. Wiley and Sons (1996).*

Interpreting Mass Spectrometer Output

The mass spectrum is a display of unique ions present at a specific time in the experiment, whether that duration represents a long-term ablation of a solid sample in the source or the passage of a transient GC or LC peak. Software is available from several sources. It is often tailored to specific practices, such as metabolite identification. It can be expeditious, reducing huge volumes of data while highlighting issues the unaided eye might overlook. Software can help us reduce uncertainty if, with properly applied skills, we make use of fundamental chemistry: the electron valence rules for nitrogen-containing compounds, characteristic spectra of halides, rings-and-double bond calculations, and so forth to arrive at what we believe is an unambiguous conclusion. No single software application can answer all inquiries satisfactorily. And so, what really counts is a practitioner's ability to apply well-honed skills and educated judgment.

A small, simple molecule such as carbon dioxide (44 Da), composed of only three atoms, produces a very simple mass spectrum. In case of CO, the molecular ion is also the most intense or abundant ion displayed (referred to as the base peak). Fragment ions found in this spectrum created from the excess internal energy of ionization are CO (m/z=28) and O (m/z=16). In some cases the molecular ion may not be the most abundant in the spectrum. For example, because cleavage of a carbon-carbon bond in propane (44 Da) gives methyl and ethyl fragments, the larger ethyl cation (m/z=29) is the most abundant. Ions derived from these well-characterized interactions are particularly significant identifying features for spectra of these hydrocarbons.

Isotope Characteristics

Since mass spectrometers separate ions by mass, distinguishing isotopes for a given element when the instrument is capable of sufficient resolution is easily accomplished. Halogenated compounds are often cited as examples, because naturally occurring bromine, for instance, consists of a nearly 50:50 mixture of isotopes having atomic masses of 79 and 81 Da. Fragmentation of Br_2 to a bromine cation then produces two equal-sized ion peaks at 79 and 81 m/z.

Even and Odd Electron Ions

Most stable organic compounds have an even number of total electrons because electrons occupy atomic orbits in pairs. When a single electron is removed from a molecule, the total electron count becomes an odd number, a radical cation. The molecular ion in a mass spectrum is always a radical cation (as seen in EI), but the fragment ions may be even-electron cations or odd-electron radical cations, depending on the neutral (uncharged) fragment lost. The simplest and most common fragmentations are bond cleavages that produce a neutral radical (odd number of electrons), and a cation having an even number of electrons. A less common fragmentation where an even-electron neutral fragment is lost produces an odd-electron radical cation fragment.

As a rule, odd-electron ions may fragment either to odd- or even-electron ions, but even-electron ions fragment only to other even-electron ions.

Mass	Odd electron ion	Even electron ion
Even	No N or even number of N atoms	Odd number of N atoms
Odd	Odd number of N atoms	No N or even number of N atoms

Table 2: The masses of molecular and fragment ions also reflect the electron count, depending on the number of nitrogen atoms in the species.

The two levels of access to interpreting mass spectra are nominal-mass data and exact-mass data. In each case, retention times serve as an additional determinant. Achieving accurate mass measurement is based on the calculated elemental composition. Not surprisingly, accurate isotope patterns fed into an algorithm to reduce the number of possible formula candidates is a recently exploited aspect of accurate mass measurement.

Characterizing Spectra Produced by Desorption and Soft Ionization

The retro Diehls Alder reactions and hemolytic/heterolytic energies required to disassociate, or cleave, bonds leading to specific well-characterized fragmentation continues to be the basis for our thinking when confronted with mass spectra. The difficult part of MS is often in answering the question posed by Fred W. McLafferty, one of the important contributors to our understanding of interpretation rules: "What is the mass we are dealing with?"

Until the development of desorption techniques like MALDI and electrospray, that question at least at times seemed easier to answer. How easy depended on whether the sample had to be derivatized to make it volatile and amenable to GC/MS. Here often the spectra would be dominated by the derivatized groups and show little or no molecular ion (hence the need for CI). In that case the advent of electrospray and APCI certainly aided in the identification of the molecular weight of small molecule singly charged species. At least in those cases MS dealt with the m/z value of ions displaying only a single charge. The mass of an analyte usually was reported as the nominal mass (the nominal m/z value) of the molecular ion, the same as the nominal mass of the molecule. The nominal mass

See MS – The Practical Art, LCGC (www.chromatographyonline.com)

Interpretation of Mass Spectra, Part I: Developing Skills, Vol. 24, No. 6, June 2006
  Why are these important: They examine how spectrometrists approach interpreting the output of mass spectrometers

Interpretation of Mass Spectra, Part II: Tools of the Trade, Vol. 24, No. 8, August 2006

Techniques for structure elucidation of unknown: finding substitute active pharmaceutical ingredients in counterfeit medicines, Vol. 25, No. 6, June 2007
  Why this is important: A thorough real world case study illustrating how MS plays a role in deducing the correct answer.

Also see

Interpretation of Mass Spectra, Fred W. McLaffferty and Franisek Turecek, University Science Books, Sausolito, CA 1993 (4th ed.)
  Why this is important: The classic text and tables for understanding the rules and approach to interpretation.

of an ion, molecule, or radical is the sum of the nominal masses of the elements in its elemental composition. The nominal mass of an element is the integer mass of the most abundant, naturally occurring, stable isotope.

The answer became more elusive when soft-ionization desorption techniques ESI became commercially widespread beginning in the early 1990s. In the "pre-desorption" age of MS, the nominal mass of most analytes interrogated by MS was less than 500 Da. Mass defect due to the presence of hydrogen was not an issue for these analytes. The upper m/z limit for most mass spectrometers fell in the 650-800 range. Thus, in those pre-desorption ionization days, the nominal mass and the integer monoisotopic mass were of the same value. The monoisotopic mass of an ion, molecule, or radical is the sum of the monoisotopic masses of the elements in its elemental composition. The monoisotopic mass of an element is the exact mass of the most abundant, naturally occurring, stable isotope.

At the onset of the desorption ionization era, larger molecules and greater precision became integral to studies because the technology permitted it with little difficulty. Only then did the issue of mass defect become so very important. In a mass spectrometer able to report only to the nearest integer m/z value, the molecular ion of a $C_{50}H_{102}$ compound might be represented by a peak at m/z 703 instead of at m/z 702 because the molecular ion would have a monoisotopic mass of 702.7825, which rounds to the integer 703.

Above 500 Da, mass defect can be a serious issue in determining the m/z values of MS peaks. It is important to keep in mind that the mass spectrometer is measuring signal intensities that occur at a specific time during the collection of the mass spectrum, regardless of the type of m/z analyzer used. The m/z value reported is a function of the time that ions of a known m/z value produced by a specific compound, relative to the calibration compound, reach the detector.

Since the mass of monoisotopic ions changes as the position on the m/z scale changes, the mass spectrometer that reports integer m/z values actually can take measurements at every 0.05 m/z units. The detected intensity can be that at the apex of the mass spectral peak or the sum of the intensities across the mass spectral peak. The m/z value reported is an integer obtained by rounding the observed m/z value for the mass spectral peak maximum.

Electron-ionization mass spectrometry often relies on perfluorinated compounds, like perfluorotributylamine (nominal molecular mass of 671), to calibrate the m/z scale. That is because the integer mass of an ion is almost the same as its monoisotopic mass. Once an ion exceeds a nominal mass of 1000 Da, there is no observed nominal m/z value peak in the mass spectrum. The monoisotopic mass peak is offset from where the nominal mass peak should be observed by an amount equal to the mass defect of the ion. For single-charge ions with masses above 500 Da, using techniques like electrospray with transmission quadrupole or quadrupole ion trap mass spectrometers that have unit resolution throughout the m/z scale, the isotope peaks will be separated clearly.

Of the many discussions on the role isotopes play in determining a compound's identity, one appeared in LCGC Europe that contributes a helpful balance. "Interpretation of Isotope Peaks in Small Molecule LC-MS" (L.M. Hill, LCGC Europe 19[4], 226-238 [2006]) is based upon low-resolution ion trap work. In a relevant part, the author cautions against overconfidence when using ion traps: "Ion trap users will have to be more careful than those with QTOF or triple quadrupole systems. It is obviously necessary to start with the +1 isotope peak isolated free of contamination . . . ion traps tend to trap with lower resolution than they scan . . . empty[ing] the trap . . . in order of mass." This does not mean ion traps cannot be used but, like all instruments, must be applied with an understanding of their abilities and their limitations.

Similarly an instrument capable of very high resolution does not automatically confer the correct answer. One data set presented in a paper by Kind and Fiehn (T. Kind and O. Fiehn, BMC Bioinformatics 7, 234 [2006]) is particularly striking and led to their conclusion, which they based on examining 1.6 million formula search results: "High mass accuracy (1 ppm) and high resolving power alone [are] not sufficient . . . only an isotopic abundance pattern filter [is] able to reduce the number of molecular formula candidates." Mass spectrometers capable of just 3 ppm mass accuracy, but 2% isotope pattern accuracy, usually remove more than 95% of the false candidates. This performance would beat even mass spectrometers capable of 0.1 ppm, if such instruments actually existed, that are not equipped with isotope pattern capability.

Between masses of 150 Da and 900 Da, the number of possible formulas listed as mass accuracy increased from 10 ppm to 0.1 ppm without the aid of isotope abundance information: from a low of 2 candidate formulas at 150 Da to 3447 at 900 Da for 10 ppm. Even at the upper end (900 Da), mass accuracy alone at 1 ppm yields 345 candidates. Invoking 2% isotope abundance accuracy, the number of candidates at 900 Da is reduced to an expedient 18. They also show that allowing a paltry 5% accuracy for isotope acquisition associated with 5 ppm accuracy yields 196 candidates.

Quantitation and Calibration

When a compound is already known, as in the case of clinical trials where statistical data from many individual samples is gathered and the administered drug and its metabolite of interest are already well characterized, full mass spectra are not required. However, very good sensitivity in a complex physiological mixture is required, so the instrument is set to monitor only specific m/z values. (Refer to the comparison of SIR and MRM response, on page 27)

Since ions continuously stream through the triple or tandem quadrupole there is no need to limit the ion current into the mass analyzer. The ion trap, on the other hand, has a defined, finite volume and, therefore, needs a function to prevent too many ions entering the trap. Not controlling ion intensity results in undesirable or unexpected peaks in a spectrum, a particularly troublesome phenomenon when attempting to library search EI spectra in GC/MS. The design of ion traps has changed substantially to allow external (ions created outside the mass filter with subsequent ion injection into the trap) rather than internal ionization (ionization within the mass filter). The design change addressed the problem of ion molecule reactions in the trap, but it also introduced a limitation. Even in MRM mode this automatic governing of the total ion current can result in irregular sampling intervals across a chromatographic peak. So, ultimately, the ion trap is limited as a tool for trace analysis in complex matrices, particularly when the highest accuracy and precision is required, as is the case, for data which must be legally defensible or whose stringent criteria for accuracy and precision of quantitation is dictated by legislation.

An internal standard is typically used when performing MS quantitation. The standard provides controls over variability in extraction processes, LC injection, and ionization. Without an internal standard RSDs among replicates can be tenfold higher than replicates referenced to a standard that typically produces RSDs in low single digits. The best internal standard is an isotopically labeled version of the molecule of interest. Although synthesizing such a molecule proves expensive, it will exhibit similar extraction recovery, chromatographic retention time, and ionization response in the mass spectrometer.

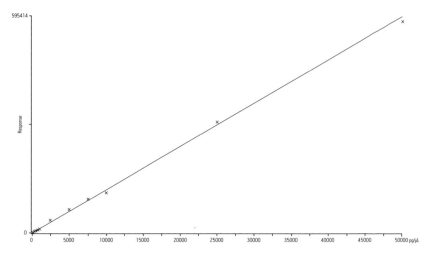

Figure 19: Quantitation example using GC/MS with CI - Linear Dynamic Range 5 Orders.

Assessing system suitability, randomizing samples, and determining suitable curves and concentration points are the subject of lengthy debate. Some good references can be seen at the www.ionsource.com tutorial on quantitation.

Calibration

Calibration compounds are used by a mass spectrometrist to adjust the mass calibration scale, as well as the relative intensities of the ions, to match that of known entities. This operation is performed on all mass spectrometers because subtle changes in electronics, cleanliness of surfaces, and a lab's ambient conditions can affect the instrument's ability to reproduce a meaningful measurement. For the least demanding analyses on nominal-mass instruments, the need for calibration can be infrequent and a check of its response more frequent. Nevertheless, high mass accuracy requires constant surveillance for minute changes.

For GC/MS, a popular calibration compound is FC-43, also known as perfluorotributylamine. Other calibration compound mixtures are used to adjust a high-resolution mass spectrometer's calibration scale. Sodium cesium iodide (NaCsI) and polyethylene glycol mixtures are popular for LCMS. In an

LCMS-compatible solvent, NaCsl streamed or infused in a steady state into an instrument provides a series of monoisotopic peaks through 4000 Da.

Kits are available that contain a selection of standard peptides, proteins, matrices, and solvents for calibrating, tuning, and sensitivity testing MALDI mass spectrometers. One, from Sigma Aldrich®, is a setup aid for analyzing complex mixtures of proteins and peptides (700 to 66,000 Da).

Lock Mass

Constant vigilance is required for the most demanding measurement with TOFs and similar highly accurate instruments. A small change in temperature alone can shift the reported mass results by many parts per million. Depending on the type of ionization used, constant calibration correction can be achieved by simply using a known contaminant present in the source. Or you can sample an ion stream periodically to re-establish proper calibration throughout an analysis. Simply adding a lock mass calibrant to flowing LC eluent, by "teeing in'" after the column and before the mass spectrometer's inlet, often causes uncontrolled behaviors like ion suppression, mass interference, and solvent effects.

TOF instruments (described on page 26) can reach low parts per million accuracy, and Fourier transform ion-cyclotron resonance (FTICR) instruments are capable of even greater accuracy, provided the number of ions admitted are well controlled. Real-time recalibration on the lock mass by corrections of any mass shift removes mass error relative to that established with calibration of the mass scale. Weak signal is another sometimes-overlooked cause, which is correctable by averaging the mass measurement over the LC peak, weighted by signal intensity.

A dual-electrospray source optimized for acquiring exact mass data is ideal for proteomics studies or low-level metabolite identification. The method using two independent ESI probes samples the correcting spray (or reference) stream, using an oscillating baffle driven by a programmable stepper motor. The reference spray is sampled at predefined intervals, ensuring the acquisition duty cycle favors the liquid stream containing the analyte. The sampling baffle position is monitored in real-time, enabling indexing of the two liquid inlets, and the reference and sample data are stored in separate files. The design eliminates cross-talk between the analyte and reference channels.

Solvents and Caveats for LCMS

Solvents are typically chosen based on a compound of interest's solubility and compatibility with various ionization techniques used in LCMS. Volatility and the solvent's ability to donate a proton are important in ESI and other atmospheric ionization techniques.

Protic primary solvents like methanol and mixtures with water, such as 1:1 methanol/water or 1:1 acetonitrile/H_2O, are used (although the water/methanol mixture increases viscosity well beyond either water or menthol as a neat solvent because of a resulting exothermic reaction). Water's relatively low vapor pressure can be detrimental to sensitivity when employed at 100%. Better sensitivity results when surface tension is decreased through addition of a volatile organic solvent. Surfactants with higher proton affinity, though they increase ion liberation from nebulized droplets, can also reduce sensitivity.

Aprotic co-solvents like 10% DMSO in water and isopropyl alcohol improve solubility for some compounds. Formic acid is often added at low levels (0.1%) to facilitate ionization by ensuring the analyte is more basic than the solvent. Even in small amounts, however, some acids, like TFA, though necessary for otherwise insoluble compounds, can limit sensitivity.

In the ESI ionization mode, buffers and salts (Na^+, K^+, and phosphate) cause a reduction in the vapor pressure and consequently a reduced signal. The increased surface tension of the droplets, and resultant reduction of volatility, can be remedied by using relatively more volatile buffers like ammonium acetate, formed by a weak acid-base pair.

Solvent considerations

- Solvent in the gas phase limits ionization by ESI to molecules more basic than the solvent. The exception is photoionization (which is not acid/base ionization) but nonetheless mediated by solvent.

- Removing solvent and water vapor from the ionization region increases types of compounds that can be ionized at atmospheric pressure.

- Reducing liquid volume relative to the sample or analyte of interest contained in the liquid improves ESI performance (i.e., lower flow rates).

- Useful solvents
    - Water
    - Acetonitrile
    - Methanol
    - Ethanol
    - Propanol
    - Isopropanol

- Acceptable additives
    - Acetic acid
    - Formic acid
    - Ammonium hydroxide
    - Ammonium formate (salt concentration = 10 mM or less)
    - Ammonium acetate (salt concentration = 10 mM or less)

- Nonvolatile salts (phosphate, borate, citrate, etc.)
    - Can deposit in source and plug capillaries thus requiring more cleaning and maintenance operations.
    - Modern source designs can handle nonvolatiles better than older designs.

- Surface-active agents (surfactants/detergents) suppress electrospray ionization

- Inorganic acids are corrosive

- Trifluoroacetic acid (TFA)
    - To some extent suppresses positive-ion electrospray at levels exceeding 0.01%.
    - Greatly suppressed negative-ion electrospray.

- Triethylamine (TEA)
    - High PA (232 Kcal/mole) yields an intense [M+H]+ ion at m/z 102.
    - Suppresses positive ion electrospray of less basic compounds.

- Tetrahydrofuran (THF)
    - 100% THF is highly flammable, so APCI and most interface techniques use nitrogen as the nebulizer gas. (Using air creates an explosion hazard.)
    - Reacts with PEEK® tubing.

Ion Suppression

Ion suppression is one of the more visible issues confronting spectrometrists using ESI as the ionization technique. The United States Food and Drug Administration's (US FDA) publication, Guidance for Industry on Bioanalytical Method Validation (Federal. Register, 66, 100, 28526) in 2001 indicates the need for such consideration to ensure the quality of analysis is not compromised. The article notes several experimental protocols for evaluating the presence of ion suppression. One compares the multiple-reaction-monitoring (MRM) response (peak areas or peak heights) of an analyte in a spiked, post-extraction sample to that of the analyte injected directly into the neat mobile phase. A low analyte signal in the matrix compared to the pure solvent indicates the presence of interfering entities.

A publication by C. Mallet et al. describes where in the chromatogram matrix effects on the analyte (and internal standard) are present. The experimenters use a continuous flow of a standard solution containing the analyte of interest and its internal standard added to the column effluent. After injecting a blank sample extract into the LC system, a drop in the constant baseline indicates suppression in ionization of the analyte due to the presence of interfering material.

Column Chemistries

One enabling change in technology is the advent of hybrid column chemistries and highly selective particles of less than two micrometers in diameter. The hybrid chemistries rely less on mobile phase modifiers that can cause ion suppression and the increased selectivity of particles.

Ultra-High Pressure LC vs. Traditional HPLC

The recent commercialization of work by Professor J. Jorgenson (University of North Carolina), often generically referred to as UHPLC (ultra-high pressure liquid chromatography), brings a potential to increase the information derived from typical LCMS analyses. Commercialized by Waters Corporation as UPLC Technology, or Ultra Performance Liquid Chromatography, the increased peak capacity relative to HPLC makes possible defining chemical entities that would otherwise have co-eluteunder the broader peaks of HPLC. Concentrating peaks into bands of (typically) two-second widths or less raises the potential for increased sensitivity by favoring the mass spectrometer's response to improvements in signal-to-noise ratio.

The UPLC Technology concept alters familiar parameters established in traditional separations practice like flow rates, particle sizes—even our appreciation of van Deemter curves. As operating pressure increases from ~2000 psi to as high as 20,000 psi, particle diameters less than 2 μm approach the theoretical limit described in 1969 by John Knox in his "Knox Equation". Once the attendant problems of increased mechanical stress and exaggerated thermal effects have been addressed the improvements in MS performance come as a somewhat counterintuitive consequence of theory.

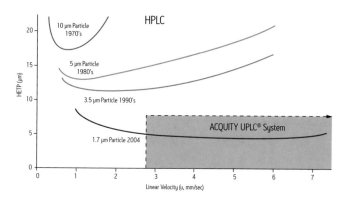

Flow Rate (mL/min)	Linear Velocity (u, mm/sec)						
i.d. = 1.0 mm	0.04	0.7	0.10	0.13	0.17	0.20	0.24
i.d. = 2.1 mm	0.15	0.3	0.45	0.6	0.75	0.9	1.05
i.d. = 4.6 mm	0.7	1.4	2.1	2.8	3.5	4.2	4.9

Figure 20: Viewed as changes in efficiency due to linear velocity depicted as a van Deemter plot, columns packed with 1.7 μm diameter particles perform better independent of flow rate.

Though all columns evidence diminished performance at extremely low linear velocities, a fact we are accustomed to in HPLC practice, those with smaller diameter particles perform better, and show less performance deterioration at increased linear velocity.

An example of how technology has redefined the approach to experimental design is seen in comparing what is now referred to as 'legacy' HPLC separations with UPLC separations. Not only have the underlying principles redefined separations (up to four times shorter) but the selectivity has increased uncovering hidden details such as the metabolites of midazolam in the figure. The improved separation indicates a second glucuronide metabolite, m/z = 548.125.

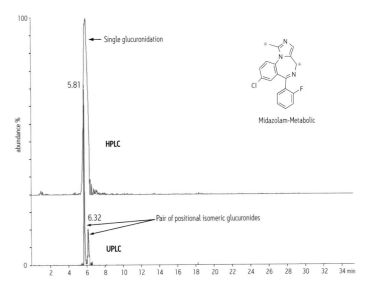

Figure 21: Technological advances often uncover more unexpected detail such as what was thought to be a single glucuronide peak.

Oxidative metabolism of midazolam is catalyzed by cytochrome P450 proteins in the liver. Two primary sites in the molecule where metabolic oxidation [hydroxylation] is most likely to occur are indicated by red asterisks in the drug's structural formula above.[1] A recent study comparing HPLC/MS with UPLC/MS for the analysis of midazolam metabolites in bile found a chromatographic HPLC peak with a nominal m/z of 548. By comparison, UPLC/MS separated a pair of peaks, each with the same exact mass value of m/z 548.1248. Observed fragmentation confirmed that these metabolites were each glucuronides. The authors stated that empirical formula consistent with the exact mass for both members of this well-separated pair could indicate that, following double hydroxylation of midazolam at the indicated sites, O-methylation could then take place at one site while the other undergoes conjugation with glucuronic acid, and vice versa.[2]

See www.waters.com/primers

[1] Kanazawa, H; et al. "Determination of midazolam and its metabolite as a probe for cytochrome P450 3A4 phenotype by liquid chromatography-mass spectrometry," J. Chromatogr. A 2004, 1031 213-218

[2] Plumb, R.; Castro-Perez, J; Granger, J; Beattie, I; Joncour, K; and Wright, A "Ultra-performance liquid chromatography coupled to quadrupole-orthogonal TOF mass spectrometry," Rapid Commun. Mass Spectrom. 2004, 18, 2331-2337

Acknowledgments

As large and diverse a task as this could not have been accomplished without the editorial overview of David Sarro who has been responsible for consistency in my writing for many years. Hilary Major provided insights and helped resolve technical questions on a variety of topics. GC and GCMS contributions by Doug Stevens and the many insights to practical LCMS by John Van Antwerp are also greatly appreciated. Finally, creating this book required ability to envision its print form and graphics, as well as the detail of its execution, to which we owe thanks to Natalie Crosier and graphics by Victoria Walton and Ian Hanslope.

Also see:

C.R. Mallet, Z. Lu, and J.R. Mazzeo, A study of ion suppression effects in electrospray ionization from mobile phase additives and solid-phase extracts, Rapid Commun. Mass Spectrom.
Vol 18, 1, 49-58, 2004
> Why this is important: Often referenced work addressing SPE comparisons.

L.L. Jessome, D. A. Volmer, Ion Suppression: A Major Concern in Mass Spectrometry, LCGC,
Vol 24, 5 May, 2006
> Why are these important: Provides a comprehensive discussion of one of the major obstacles encountered in ESI gas-phase proton affinity changes affecting quantitation.

U. Neue, Theory of peak capacity in gradient elution, J Chromatogr A. 2005 Jun 24;1079 (1-2):153-61
> Why this is important: Contemporary discussion of modern chromatography practice.

Toyo'oka, Toshimasa, Determination Methods for Biologically Active Compounds by Ultra-Performance Liquid Chromatography Coupled with Mass Spectrometry: Application to the Analyses of Pharmaceuticals, Foods, Plants, Environments, Metabonomics, and Metabolomics, Journal of Chromatographic Science, Vol 46, 3, March 2008, 233-247
> Why this is important: Provides a comprehensive overview of publications related to modern chromatography practice.

J. H. Knox and M. Saleem, Kinetic Conditions for Optimum Speed and Resolution in Column Chromatography, J. Chromatogr. Sci., 7 (1969), 614-622
> Why this is important: One of the fundamental papers predicting modern chromatography practice.

Glossary

The following list of terms is derived from common usage throughout the industry as an adjunct to the discussions in this primer and includes terms and techniques no longer in common usage.

Abundance

When viewed as similar to absorbance displayed on a UV detector the vertical increase in signal above background indicates an increased occurrence of that particular ion (when the x axis is calibrated in mass units) or total ions present (when the horizontal axis is calibrated in time or scans). The signal for all ions resulting from the fragmentation of a single analyte or compounds compared to a base peak (the relative abundance of each ion) is used to determine the fit of a fragmented pattern to a library spectrum for positive identification.

Accurate Mass

The measured mass value for a compound with an associated error like 5 ppm. Accurate mass is also commonly used to refer to the technique rather than the measured mass. Exact mass is the exact theoretical value for the mass of a compound.

Atmospheric Solids Analysis Probe (ASAP)

Based on work by Horning in the 1970s, this form of sample ionization, developed by McEwen and McKay, uses a standard APCI plasma but forms ions by placing the sample in a heated nitrogen stream. The heat volatilizes a surprisingly large number of samples and ions are formed by charge exchange with metastable ions created by the APCI plasma. Relatively unambiguous identifications can be made of individual compounds from complex mixtures at low levels using accurate mass instruments. See also DART and DESI.

Atmospheric Pressure Ionization (API)

The term used to refer generally to techniques such as electrospray ionization (ESI) and atmospheric pressure chemical ionization (APCI) and others that operate at atmospheric pressure.

Atmospheric Pressure Chemical Ionization (APCI)
Originally called solvent-mediated electrospray, it is more often successfully applied to neutral molecules that don't ionize easily directly out of solution. APCI provides a current on a sharp pin, positioned in the on coming aerosol stream, to create a plasma of metastable ions from the solvent itself and transfer the charge from these ions to the analyte as it passes through the plasma. Heating a probe through which the LC or solvent stream passes creates the aerosol.

Atmospheric Gas Chromatography
Developed by Charles McEwen at DuPont in 2002. Using a heated transfer line a standard GC effluent can be introduced to a standard API (or ESI/APCI) source on a mass spectrometer. This provides an easy and fast change over from ESI to GC for compounds that would be best analyzed by GC. Mode of ionization can be either APCI or APPI.

Atmospheric Pressure Photoionization (APPI)
Developed in the 1980s, but commercialized after 2000 when krypton gas lamps were found to generate sufficient photon energy at 10 eV (approximately) to ionize non-polar analytes,t such as PAHs and steroids not typically amenable to ESI and APCI ionization.

Base Peak
Usually the most intense peak in the spectrum to which others are compared; in ionization techniques which give extensive structural information such as EI the base peak may not be the parent or molecular ion.

Calibration
Substances of known mass are introduced usually as a constant flowing stream while the mass spectrometer software acquires a signal for a given set of filtering conditions (i.e., RF/DC ratio for a quadrupole instrument). After comparing the acquired signal to a reference file, a calibration look-up table is created in the software. The calibration table is then the basis for the mass-to-charge ratios passed by the quads to be assigned a specific value. See sections on 'Quantitation and Calibration"

Charge-Residue Mechanism

Related to electrospray ionization; a theory first proposed in 1968 by Malcolm Dole in which he hypothesized that as a droplet evaporates its charge remains unchanged. The droplet's surface tension, ultimately unable to oppose the repulsive forces from the imposed charge, explodes into many smaller droplets. These Coulombic fissions occur until droplets containing a single analyte ion remain. As the solvent evaporates from the last droplet in the reduction series, a gas-phase ion forms.

Chemical Ionization (CI)

Collisions induced at low vacuum (0.4 torr) by the introduction of a reagent for the purpose of enhancing the production of molecular ions and often sensitivity; as this is a much lower energy process than electron impact ionization fragmentation is reduced and it is often referred to as a soft ionization technique. See Electron Ionization.

Collision-Induced Dissociation (CID)

Also referred to as collisionally-activated dissociation (CAD), is a mechanism by which molecular ions are fragmented in the gas phase by acceleration (using electrical potential) to a high kinetic energy in the vacuum region followed by collision with neutral gas molecules, such as helium, nitrogen, or argon. A portion of the kinetic energy is converted or internalized by the collision which results in chemical bonds breaking and the molecular ion is reduced to smaller fragments. Some similar 'special purpose' fragmentation methods include electron-transfer dissociation (ETD), electron-capture dissociation (ECD). See the section on 'Bio-molecular ionization methods'.

Direct Analysis Real Time (DART)

Developed by Robert Cody and others in 2002, similar to DESI in application although more closely related to APCI in function. A sample is placed on a substrate and bombarded by energized particles formed in a process similar to APCI. That is, metastable ions are formed by a plasma and transported by heated nitrogen gas directed at the target. See also McEwen's work listed under Atmospheric GC and ASAP.

Delayed Extraction (DE)

Developed for MALDI-TOF instruments, "cools" and focuses the ions for approximately 150 nanoseconds after they form and before accelerating the ions into the flight tube. The cooled ions have a lower kinetic-energy distribution than uncooled ones, and they ultimately reduce the temporal spread of the ions as they enter the TOF analyzer, resulting in increased resolution and accuracy. DE is significantly less advantageous with macromolecules (for instance, proteins >30,000 Da).

Desorption-Electrospray Ionization (DESI)

First described by Graham Cooks in 2002 as a means of producing soft secondary ions from (typically) an inert substrate surface. Analogous to MALDI using an ESI probe aimed at about 50o incident angle to the surface allowing ions to chemically sputter and be admitted to the mass spectrometer. Shown to produce information directly from many polar and non-polar surface materials (skin, intact fruit for pesticide residue detection, etc) without the need for sample preparation. See also McEwen's work listed under Atmospheric GC and ASAP.

Desorption Ionization on Silica (DIOS)

Once viewed as an alternative to a preparing samples in a MALDI substrate especially for small molecules since the substrate (a bare silica surface) would not generate interfering ions. Its commercial potential in the late 1990s diminished as the difficulties of producing the plates and the surface susceptibility to contamination became apparent.

Direct Current

For our interests the term is usually used in conjunction with 'radio frequency' when describing how the quadrupole functions as a mass filter. Superimposed radio frequency (RF) and constant direct current (DC) potentials between four parallel rods were shown by Wolfgang Paul in 1953 to act as a mass separator, or filter, where only ions within a particular mass range, exhibiting oscillations of constant amplitude, could collect at the analyzer.

Electron Ionization (EI)

Sometimes incorrectly referred to as "electron impact" ionization resulting from the interaction of an electron with a particle (atom or molecule); can be thought of as a 'hard' ionization technique since sufficient energy is imparted to disrupt internal chemical bonds requiring high kcal/mol. Ionizing voltage (typically 70eV) refers to the difference in voltage causing acceleration of the electrons used to induce electron ionization. Unlike CI, EI avoids uncontrolled collisions by operating at high vacuum. The analyzer operates at even higher vacuum (10^{-4} to 10^{-6} torr).

Electrospray Ionization (ESI)

A so-called 'soft' ionization technique. The most widely employed of the atmospheric-pressure ionization (API) techniques. Commercially significant since the late 1980s, the phenomenon is attributed to an excess of energy (voltages in the 3-5 kV range) applied to a conductive tube (stainless steel capillary) inducing the liquid flowing inside to exceed its rayleigh limits and forming an aerosol upon exiting the tube. The resulting spray (the result of a coulombic explosion) then gives rise to ions contained in the aerosol droplets as they desolvate to approximately 10 micron radius. The ions are typically protonated and detected in the form M+H in positive ionization mode or M-H in negative ion mode.

Elemental Analysis

The nominal mass of an ion, molecule, or radical is the sum of the nominal masses of the elements in its elemental composition. Achieving accurate mass measurement is based on the calculated elemental composition but the term 'elemental analysis' is typically performed on inorganic materials—to determine elemental makeup, not structure—in some cases using solid metal samples. Inductively coupled plasma (ICP) sources are common where a discharge (or lower power-glow discharge) device ionizes the sample. Detection using dedicated instruments, at the parts-per-trillion level, is not uncommon

Exact Mass

The exact theoretical value for the mass of a compound. 'Accurate Mass' is the measured mass value for a compound with an associated error like 5 ppm. Accurate Mass is also commonly used to refer to the technique rather than the measured mass.

Fast Atom Bombardment (FAB)

One of the earlier so-called soft-ionization techniques where the result is usually an intense molecular ion with little fragmentation. The analyte is placed in a matrix (often glycerol) either flowing or more commonly placed on the tip of a probe and positioned in the path of high energy atoms—often xenon or cesium iodide. The technique has been effective for biomolecules up to 10,000 amu but perhaps more important in conjunction with the magnetic sector mass spectrometer where exact weight can be determined such as for novel peptides. Sensitivity can be very good (low femtomol levels). The technique can be difficult to master, the glycerol does foul the mass spectrometer source and the lower masses can be obscured by the presence of glycerol ions. This technique is little used now since the introduction of ESI.

Field Ionization (FI)

The soft ionization provided by FI leads to little or no fragmentation for a wide variety of analytes. This is particularly significant in petrochemical applications where there are limitations to the utility of other ionization techniques due to fragmentation (EI) or complex ionization characteristics (CI). A thin wire with a high applied voltage is heated in the vapor of an organic compound such as indene. The resulting structures which are deposited on the surface of the wire, dendrites, are pyrolized producing very fine conductive filaments. When a very fine point has a high potential applied to it a very intense electric field is generated at this tip providing conditions under which field ionization can take place.

Sample molecules pass in close proximity to the tips of a mass of carbon dendrites grown on the FI emitter. The FI emitter is positioned in close proximity to a pair of hollow extraction rods. The emitter is held at ground potential and a relatively high voltage (12 kV) is applied to the rods producing very high electric fields around the tips of the carbon dendrites. The GC column is positioned in close proximity and in line with the emitter wire. Under the influence of the electric fields, quantum tunneling of a valence electron from the molecule takes place to give an ion radical.

Flow Injection Analysis (FIA)
This is the practice of introducing a sample (usually purified in an earlier step, as a fraction to remove interferences and complexity from the resulting spectrum) through the LC injector but without a column in-line. The LC acts as a sample introduction device only.

Filament
In electron ionization the filament is the source of electrons which interact with the analyte to ionize it. Typically made from a metal wire (flat or round) capable of giving off 70 eV electrons as it heats from current passing through it.

Fragment Ion
An ion produced as the loss from a parent molecular ion. The sum of the dissociated fragments equals the parent and under given conditions will always fragment the same internal bonds to produce a predictable pattern (same ions and relative abundances for each). See also Product (daughter) ions resulting from specific MRM experiments.

Gas-Phase Ion
Analytes must be converted from their resting state to an ion in order to be manipulated and acquired by a mass spectrometer. There are a number of ways to accomplish this as described in the primer—some more aggressively creating fragments while others conserve the analyte intact. Energy is presented to the analyte producing an ion in the gas phase as opposed to, for instance, the condensed phase used to separate analytes in LC.

Hybrid
Usually refers to an instrument that is a combination of two different types for instance, earlier 'hybrids' combined magnetic sector and quadrupoles. Today's QTOF is a hybrid of quadrupole and TOF.

Ion
Mass spectrometers can only manipulate and therefore detect a mass when it possesses at least one charge. When only one charge is present (for instance by the loss of an electron causing the molecule to exist as a positively charged cation radical or by the addition of a proton or hydrogen to exist as a positively charged pseudo-molecular ion) we can think of it as representing the molecular weight in a low-resolution scheme.

Ion Current (Total Ion Current)
The electric current detected based on the charged particles created in the ion source. If the mass spectrometer is set to scan over a range of 100 to 500 Da the resulting total ion current will be the sum of all ions present in the source within that range at the selected time. If the instrument is set to detect only one ion (selected ion monitoring) the resulting total ion current will be the sum of only that ion at each selected instance.

Ion Cyclotron (ICR)
Cyclotron instruments trap ions electrostatically in a cell using a constant magnetic field. Pulses of RF voltage create orbital ionic motion, and the orbiting ions generate a small signal at the detection plates of the cell (the ion's orbital frequency). The frequency is inversely related to the ions' m/z, and the signal intensity is proportional to the number of ions of the same m/z in the cell. At very low cell pressures, a cyclotron instrument can maintain an ion's orbit can for extended periods providing very high resolution measurements. Fast fourier transform ion cyclotrons (FTICR) represent the extreme capability of measuring mass with the ability to resolve closely related masses. Although impractical for most applications, a 14.5-tesla magnet can achieve a resolution of more than 3.5 million and thus display the difference between molecular entities whose masses vary by less than the mass of a single electron.

Ion-Evaporation Mechanism

Reference to electrospray ionization; in 1976, Iribarne and Thomson proposed the ion evaporation mechanism (see also 'charge-residue mechanism') in which small droplets form by Coulombic fission, similar to the way they form in Dole's charge residue model. However, according to ion-evaporation theory, the electric field strength at the surface of the droplet is high enough to make leaving the droplet surface and transferring directly into the gas phase energetically favorable for solvated ions.

Ion Mobility

Is a technique that differentiates ions based on a combination of factors: their size, shape, charge, and mass. Ion mobility measurements and separations coupled with tandem mass spectrometry can help overcome analytical challenges that could not be address by other analytical means, including conventional mass spectrometry or liquid chromatography instrumentation. Ion mobility mass spectrometry (or IMMS since 'imaging mass spectrometry' is often abbreviated 'IMS') devices are commonly used in airports and hand-held field units for rapidly (20 msec) detecting small molecules whose mobility is known. A wide range of analytes are amenable to IMMS.

Ion Source

The physical space in the ion stream in front of the analyzer where the analyte is ionized. Each type of interface requires its own internal geometry for optimum results.

Ion Trap

Also linear trap and Q-trap (or quadrupole trap). Wolfgang Paul's invention of the quadrupole and quadrupole ion trap earned him the Nobel Prize in physics in 1953. An ion trap instrument operates on principles similar to those of a quadrupole instrument. Unlike the quadrupole instrument, however, which filters streaming ions the trap stores ions in a three-dimensional space. Before saturation occurs from too many ions attempting to fill the finite space, the trap or cyclotron allows selected ions to be ejected for detection. Fields generated by RF voltages applied to a stacked or "sandwich" geometry (end-cap electrodes at opposing ends) trap ions in space between the two electrodes. Ramping or scanning the RF voltage ejects ions from their secular frequency, or trapped condition. Dynamic range is sometimes limited. The finite volume and capacity for ions limits the instrument's range, especially for samples in complex matrices.

The ability to perform sequential fragmentation and thus derive more structural information from a single analyte (i.e., fragmenting an ion, selecting a particular fragment, and repeating the process) is called MSn. GC chromatographic peaks are not wide enough to allow more than a single fragmentation (MS/MS or MS^2). Ion trap instruments perform MS/MS or fragmentation experiments in time rather than in space, like quadrupole and sector instruments. So they cannot be used in certain MS/MS experiments like neutral loss and precursor ion comparisons. Also, in MS/MS operation with an ion trap instrument, the bottom third of the MS/MS spectrum is lost, a consequence of trap design. To counter the loss, some manufacturers make available via their software wider scan requirements that necessitate the switching of operating parameters during data acquisition.

Because of similarities in functional design, quadrupole instruments are hybridized to incorporate the advantages of streaming quadrupole and ion trapping behavior to improve sensitivity and allow on-the-fly experiments not possible with either alone. Such instruments are sometimes called linear traps or Q-traps). The increased volume of a linear trap instrument (over a three-dimensional ion trap) improves dynamic range.

Isotope Ratio
Although often presumed to be constant and stable, natural isotope abundance ratios show significant and characteristic variations when measured very precisely. Isotope ratio measurements are useful in a wide range of applications, for example, metabolic studies using isotopically enriched elements as tracers; climate studies using measurements of temperature-dependent oxygen and carbon isotope ratios in foraminifers; rock age dating using radiogenic isotopes of elements such as lead, neodymium, or strontium; and source determinations using carbon isotope ratios (to determine if a substance is naturally occurring or is a petroleum-based synthetic).

Typically, single focusing magnetic sector mass spectrometers with fixed multiple detectors (one per isotope) are used. Complex compounds are reduced to simple molecules prior to measurement, for example, organic compounds are combusted to CO_2, H_2O and N_2.

Laser Ablation
Compounds can be dissolved in a material which acts as an intermediate to transfer charge to the analytes of interest. A laser is aimed at the mixture to cause sputtering to produce ions in the space just above the mixture where they can be sampled or drawn into the mass spectrometer. Especially useful as a very soft ionization technique to look at intact large molecules since the low mass ions from the matrix produce a highly complex, intense background which could otherwise interfere with analytes of similar low mass.

Magnetic Sector

Ions leaving the ion source are accelerated to a high velocity; they then pass through a magnetic field perpendicular to their direction. When acceleration is applied perpendicular to the direction of motion, the object's velocity remains constant, but the object travels in a circular path. Therefore, the magnetic sector is designed as an arc. Ions with a constant kinetic energy, but different mass-to-charge ratio are brought into focus at the detector slit (called the 'collector slit'') at different magnetic field strengths.

The magnetic sector alone will separate ions according to their mass-to-charge ratio. However, since the ions leaving the ion source do not have exactly the same energy (therefore, do not have exactly the same velocity) the resolution will be limited. An electric sector that focuses ions according to their kinetic energy was usually added which like the magnetic sector applies a perpendicular force to the direction of ion motion.

Matrix-Assisted Laser Desorption Ionization (MALDI)

First introduced in 1988 by Tanaka, Karas, and Hillenkamp, uses a laser to strike and energize a matrix containing the analyte. It has proven to be the method of choice for ionizing exceptionally large peptide and protein molecules that can then be detected intact. Commonly employed as the introduction scheme for time-of-flight (TOF) instruments often referred to as MALDI-TOF.

Mass-to-Charge Ratio (m/z)

Charged particles are represented as a ratio of their mass to their ionic charge. In literature and general use this often appears as 'm/z' where the analyte from which the ion is derived might be labeled using atomic mass units (amu), daltons or molecular weight (mw).

Mean Free Path

The distance from entrance of an ion into the analyzer and detection of that ion. At operating vacuum the mean free path is relatively long considering time between collisions in rarified air versus the time needed to analysis an ion. Example:

$$\text{atm (1000 torr) air contains } 3 \times 10^{22} \text{ molecules/cm}$$
$$\text{chamber at } 1 \times 10^{-5} \text{ torr contains } 3 \times 10^{11} \text{ molecules/cm}$$
$$\lambda \div \text{pressure (torr)} = \text{min. mean free path (cm)}$$
$$\text{where } \lambda = 5 \times 10^{-3} \text{ cm}$$

Molecular Ion
The ion produced when a molecule gains (anion) or losses (cation) an electron. See also Pseudo-molecular ion.

Monoisotopic Mass
The monoisotopic mass of an element is the exact mass of the most abundant, naturally occurring, stable isotope. In a mass spectrometer able to report only to the nearest integer value, the molecular ion of a $C_{50}H_{102}$ compound might be represented by a peak at m/z 703 instead of at m/z 702 because the molecular ion would have a monoisotopic mass of 702.7825, which rounds to the integer 703. Once an ion exceeds a nominal mass of 1000 Da, there is no observed nominal m/z value peak in the mass spectrum. The monoisotopic mass peak is offset from where the nominal mass peak should be observed by an amount equal to the mass defect of the ion. For single-charge ions with masses above 500 Da, using techniques such as electrospray with transmission quadrupole or quadrupole ion trap mass spectrometers that have unit resolution throughout the m/z scale, the isotope peaks will be separated clearly. See Nominal Mass.

Multiple Reaction Monitoring (MRM)
A specific experiment on a triple quad mass spectrometer where a parent ion is filtered in the first quad (Q1), a collision is then induced between the parent ion and a molecule (usually a gas such as argon) in the middle or 'RF only' quad (Q2) followed by detection of a specific product ion from that collision (Q3). Used in high-throughput quantitative analyses in the pharmaceutical industry especially.

MSn
A term coined with the resurgence of ion traps denoting the ability to choose a specific ion present in the ion source and fragment it; repeating the procedure to increase specificity when attempting to iden-tify an analyte. The procedure can be repeated (number of repeats = n) providing the chosen fragments have sufficient energy and enough sample and time are provided for the experiment to continue.

Nominal Mass
Electron ionization MS often relies on perfluorinated compounds like perfluorotributylamine (nomi-nal molecular mass of 671) to calibrate the m/z scale. That is because the integer mass of an ion is almost the same as its monoisotopic mass. Once an ion exceeds a nominal mass of 1000 Da, there is no observed nominal m/z value peak in the mass spectrum. The monoisotopic mass peak is offset from where the nominal mass peak should be observed by an amount equal to the mass defect

of the ion. For single-charge ions with masses above 500 Da, using techniques like electrospray with transmission quadrupole or quadrupole ion trap mass spectrometers that have unit resolution throughout the m/z scale, the isotope peaks will be separated clearly. See Monoisotopic Mass.

Open Access (OA)
Also referred to as 'walk-up systems' are workflow controls allowing a fully trained operator to create complete LC or GC/MS methods and make them available to a large number of non-specialist users giving them access to advanced technology without the requirement for extensive training.

Parent Ion
More properly referred to as 'precursor'; a generally interchangeable term with 'molecular ion'. Use of this term infers the presence of a product ion in an MRM scheme. See Product Ion.

Particle Beam (MAGIC, ThermaBeam®)
Originally developed at Georgia Tech and dubbed MAGIC (monodisperse aerosol generating interface for chromatography) by Browner, et al. The technique was later refined and is generically referred to as particle beam. The LC stream is heated and nebulized to remove the solvent. Vacuum pumps draw the solvent vapor through skimmer cones in series (usually two). The result is a "dried" particle that accelerates through the momentum separator and impacts the mass spectrometer source producing fragment ion spectra similar to traditional GC/MS. This technique is another which is now rarely used.

Probe, also Solids Probe or Direct Insertion Probe
A metal rod inserted into the mass spectrometer source through a vacuum lock. Samples can be applied to the tip of the probe and placed into the path of an ionizing beam. Typically used for EI and other single sample manual experiments. Samples can also be applied in an ionization enhancing matrix as in the case of FAB. See FAB.

Product Ion
Formerly called "daughter ions" these are the result of controlled experiments where a precursor (or 'parent') ion and molecule collisions are intentionally induced to cause fragmentation. The collision gives rise to a product ion specific to the precursor ion and is used as a means of positive identification. See Multiple Reaction Monitoring (MRM).

Protonated Molecular Ion

Some forms of ionization produce ions by a proton transfer process that preserves and promotes the appearance of the molecular ion itself (the end result referred to as a Pseudo-molecular ion). In chemical ionization, for instance, the sample is exposed to an excess of reagent gas to form a protonated molecular ion (represented M+H). The reverse process can produce negative ions. Transferring the proton to the gas molecule can, in some cases, produce the negative ion (M-H) or deprotonated ion.

Pseudo-Molecular Ion

Usually refers to the adduction of a proton (e.g., M+H) or ion (e.g., $M+NH_4$ derived from the ammonium salt commonly used in the mobile phase) that alters the analyte of interest in some relatively easily identifiable fashion. The charge allows manipulation by the mass spectrometer.

Quadrupole (Quad)

The underlying feature for the most prevalent type of mass spectrometer. Four rods (often no more than 1" in diameter and less than 12" long) are held parallel to each other (about 1" apart) in two collars. Filtering, or passing a given charged particle along its length, is accomplished by applying direct current (DC) and radio frequency (RF) voltage to the rods. Different masses (with associated charge) are affected by changing the RF/DC conditions. The rods are connected as paired opposites— each set alternated as the positive and negative poles by the RF source.

For a given calibrated setting, particles of corresponding mass-to-charge ratio only will pass (represented as m/z; approximately equivalent to molecular weight). The same setting will cause higher weight particles to miss the detector by passing in oblique fashion to the poles (the voltage settings having little or no effect) and lighter particles to become entrapped without reaching the exit and being detected. Quadrupoles can change and stabilize these mass filter field conditions quickly allowing more than one molecular weight to be observed by scanning over time although fewer charged particles are therefore detected for any given molecular weight.

Functionally the single quadrupole mass filter, used alone when matrix interference is not an issue, can be joined with another to enhance discrimination of a given analyte among many in a background (from the matrix for instance). Early designs used a third quadrupole-type device between the two as the collision cell (hence the term 'triple quadrupole') while more recent designs use specialized devices and are referred to as 'tandem quadrupoles'.

Quadrupole Time-of-Flight (QTOF)

This mass spectrometer couples a time-of-flight (TOF) instrument with a quadrupole instrument. This pairing results in the best combination of several performance characteristics: accurate mass measurement with a TOF and the ability to carry out fragmentation experiments between the two along with high quality quantitation and mass filtering. A QTOF instrument's high mass accuracy falls within a few parts per million of the true, calculated, monoisotopic value, and its high resolution—as much as 10 times higher than a quadrupole instrument's—permits us to determine empirical formulas according to mass defect (where the critical mass value of hydrogen and other atoms present serve as a differentiator).

Radical Cation

Most stable organic compounds have an even number of total electrons because electrons occupy atomic orbits in pairs. When a single electron is removed from a molecule, the total electron count becomes an odd number, a radical cation. The molecular ion in a mass spectrum is always a radical cation (as seen in EI), but the fragment ions may be even-electron cations or odd-electron radical cations, depending on the neutral (uncharged) fragment lost. The simplest and most common fragmentations are bond cleavages that produce a neutral radical (odd number of electrons), and a cation having an even number of electrons. A less common fragmentation where an even-electron neutral fragment is lost produces an odd-electron radical cation fragment.

Radio Frequency – see 'Direct Current'

Resolution (10% valley method)

The minimum separation between two neighboring masses of approximately equal response for the mass spectrometer to distinguish between ions of different mass-to-charge ratio. More typically used with magnetic sectors; equal to the ratio of:

$$\frac{\text{average mass of the two particles}}{\text{difference in their masses}}$$

Resolution (M/ΔM)

More commonly used as a measure where a given mass is divided by the resolution at full width half height maximum (FWHM). The 10% valley method was prevalent with magnetic sector instruments and requires the neighboring masses be of equal intensity. For instance, a typical resolution value for a quadrupole is 0.6 amu at FWHM. Measured using an acquired peak at mass-to-charge 3000 (equivalent to Daltons or amu) equals a resolution of 5000. The results of the two techniques are roughly comparable with this method typically yielding values double the valley method.

$$\frac{\text{mass}}{\text{peak width at 50\% peak height}}$$

Root-Mean Square Error Measurement (RMS)

A comprehensive method of evaluating instrument mass accuracy measurement capability which resembles intended use is by calculating the root mean square or RMS error. The RMS error is calculated using the following relation, where E_{ppm} is the ppm error, and n is the number of masses considered:

$$RMS = \sqrt{\frac{\sum (E_{ppm})^2}{n}}$$

Scanning

Control voltages (DC and RF) are adjusted by the computer over a given time to scan (detect) any charged particles in the specified range. The benefit of being able to detect more than one species is at the expense of sensitivity since some of the desired particles will undoubtedly be available to the detector while it is set to detect elsewhere in the range. See Selected Ion Monitoring, Quadrupole and Ion Current.

Selected Ion Monitoring (SIM)

Also called Selected Ion Recording (SIR); refer also to quadrupole and scanning. The DC and RF voltage settings on the quadrupoles can be adjusted to pass only one charged particle (a single mass-to-charge ratio) through to the detector. The result is a dramatic decrease in noise allowing the signal to appear as a dramatic increase in sensitivity (all particles of that m/z are being detected all the time) at the expense of any other particles in the mixture being detected at all.

Thermospray

Although this type of interface has been in the literature for some time, it was popularized in the early 1980s. Vestal and Blakely should be given credit for creating the first true commercially feasible interface between LC and MS. LC solvent at approximately 1 mL/min is heated in a probe (insulated tubing approximately 1-2 feet long and 75-150 microns internal diameter) and the resulting vapor is sprayed into the mass spectrometer. Ions created by the desolvation of the aerosol droplets inside the mass spectrometer enter the analyzer (at right angles to the spray) and are affected by the lens voltages. See Scanning, Ion Current.

The spectra produced are termed soft ionization spectra since little meaningful fragmentation is produced. An intense molecular ion is produced and while the single ion may be of little advantage in some backgrounds and mixtures it is advantageous for high molecular weight confirmation at great sensitivity and for filtering a target ion for further fragmentation (MS/MS). Literature reports low pmol sensitivity for vitamin D metabolites. The interface was generally chosen for highly polar applications such as metabolite work before the refinement of APCI in the early 1990s and worked poorly as organic content in the liquid increased.

Time-of-Flight (TOF) Mass Spectrometer
A mass analyzer that separates ions of different mass-to-charge ratios by their time of travel through a field-free vacuum region after having been given the same kinetic energy. The velocity of the ions is dependent on their mass-to-charge ratio and as the ions are traveling over a fixed distance the time taken to reach the detector allows the mass-to-charge ratios to be determined with heavier ions taking longer.

Tuning
Typically refers to optimizing the interface lenses and flowing gases to achieve a desired response for a specific analyte under a set of operating conditions as opposed to calibration. Calibration defines the mass acquisition and reporting function. Hardware settings of lenses and related circuits in conjunction with creation of a software lookup table sets a stable instrument's response, corrected to a list of known masses from a flowing stream of calibrant such as PEG or NaCsI.

Vacuum (torr)
Equivalent to 1 mmHg (1 psi = 51.7 torr = 0.069 bar or atm). The analyzer portion of the mass spectrometer typically must be maintained at a minimum of 10^{-4} torr to allow discrete passage of the ionized particles. Pressures tending toward atmospheric cause ion-molecular interactions which can produce random results in the charged particles being detected further downstream. Under controlled conditions such collisions are induced at low vacuum (higher pressure) for techniques, such as chemical ionization (CI).